STELLINGEN

I

Het mechanisme van Strojny *et al.* voor de oxidatie van 2-methoxyethanol met behulp van salpeterzure oplossingen geeft geen verklaring voor de grote specificiteit van deze reactie, waarbij 2-methoxyazijnzuur bijna kwantitatief wordt gevormd.

E. J. Strojny, R. T. Iwamasa and L. K. Frevel, *J. Amer. Chem. Soc.* **93**, 1171 (1971).

II

Aan een ander mechanisme dan dat van Astley en Sutcliffe voor de vorming van α,α-dichloor-*N*-trichloormethylnitron bij thermolyse van trichloornitrosomethaan dient de voorkeur te worden gegeven.

V. Astley and H. Sutcliffe, *Tetrahedron Letters* 2707 (1971).

III

De opmerking van Barashkin *et al.* dat de reactiviteit van *m*-nitrofenol voor reactie met γ-butyrolacton verhoogd wordt door de aanwezigheid van fenolaat-anionen is ongegrond.

V. A. Barashkin, N. A. Aliev and Ch. Sh. Kadyrov, *J. Org. Chem. USSR* **6**, 2274 (1970).

IV

Niet alleen de "inconsistent results" van Mutai (betreffende de intramoleculaire charge-transfer interactie tussen *p*-nitrophenoxy- en arylaminogroepen), maar vooral ook de desondanks daaraan gekoppelde beschouwingen zijn twijfelachtig te noemen.

K. Mutai, *Tetrahedron Letters* 1125 (1971).

V

De conclusie van Kingston *et al.* dat molecuulionen van cyclische acetalen soms langs misleidende routes kunnen fragmenteren, is ongegrond aangezien het massaspectrum van het betreffende geacetaliseerde derivaat van *flourensic acid* ook verklaard kan worden met een normale fragmentatie-route.

D. G. I. Kingston, M. M. Rao and T. D. Spittler, *Tetra‑ hedron Letters* 1613 (1971).

VI

De satellietbanden in het laser-Raman spectrum van koolstofsuboxide in de gasfase zijn ten onrechte door Smith en Barrett geïnterpreteerd als zijnde de "hot bands" $v_s + (n - n)v_7$.

W. H. Smith and J. J. Barrett, *J. Chem. Phys.* **51**, 1475 (1969).

VII

Omdat het mechanisme van Lewandos en Pettit voor de door metaal ge- katalyseerde disproportionering van olefinen alleen maar een enigszins andere voorstelling is van een reeds bekend mechanisme, dient aan het nut van hun artikel getwijfeld te worden.

G. S. Lewandos and R. Pettit, *Tetrahedron Letters* 789 (1971).
F. D. Mango and J. H. Schachtschneider, *J. Amer. Chem. Soc.* **89**, 2484 (1967).

VIII

Hoewel de chemiluminescentie-reactie van violanthron zeker in het *Journal of Chemical Education* thuis hoort, kan dat van het door Slabaugh gegeven "mechanisme" bepaald niet gezegd worden.

W. H. Slabaugh, *J. Chem. Educ.* **47**, 522 (1970).

IX

Het is zeer onwaarschijnlijk dat de enorme toename van de sterfte ten gevolge van longkanker bij mannen in de jaren 1950–1969 uitsluitend toe- geschreven zou moeten worden aan veranderde rookgewoonten.

Statistiek van doodsoorzaken (*CBS*), bewerkt door de Ge- neeskundige Hoofdinspectie van de Volksgezondheid.
Statistieken van de nederlandse sigaretten- en shagproductie en van de sigarettenimport (*CBS*).

J. W. Hartgerink   Amsterdam, 8 december 1971

SPIN TRAPPING BY NITROSOALKANES

MECHANISMS OF SOME PHOTOCHEMICALLY INDUCED REACTIONS

SPIN TRAPPING
BY NITROSOALKANES

MECHANISMS OF SOME
PHOTOCHEMICALLY INDUCED REACTIONS

ACADEMISCH PROEFSCHRIFT

*ter verkrijging van de graad van doctor
in de wiskunde en natuurwetenschappen
aan de Universiteit van Amsterdam
op gezag van de rector magnificus
in het openbaar te verdedigen in de aula der universiteit
(tijdelijk in de Lutherse kerk, ingang Singel 411, hoek Spui)
op woensdag 8 december 1971 des namiddags te 3 uur precies*

door

JAN WILLEM HARTGERINK

geboren te Voorburg

SPRINGER-SCIENCE+BUSINESS MEDIA, B.V.
1971

Promotor: PROF. DR. TH. J. DE BOER

Coreferent: DR. J. B. F. N. ENGBERTS

Aan mijn ouders

VOORWOORD

Bij de voltooiing van dit proefschrift wil ik gaarne mijn dank betuigen aan allen die hebben bijgedragen aan mijn wetenschappelijke vorming en vooral ook aan die medewerkers van het Laboratorium voor Organische Scheikunde, die medewerking hebben verleend aan het tot stand komen van deze dissertatie.

Hooggeleerde de Boer, U ben ik zeer erkentelijk voor uw stimulerende leiding en voor de waardevolle kritiek die ik vooral ook bij het schrijven van de artikelen, waarop de inhoud van dit proefschrift mede berust, heb mogen ontvangen.

Zeergeleerde Engberts, beste Jan, ook na je "Amsterdamse jaar" ben je het onderzoek vanuit Groningen op stimulerende wijze blijven steunen. Hiervoor en voor je werkzaamheden als coreferent ben ik je veel dank verschuldigd.

De Heren L. C. J. van der Laan en P. W. Baas, alsmede Drs. J. J. Zeilstra dank ik voor het enthousiasme waarmee zij een bijdrage hebben geleverd aan het experimentele gedeelte van het onderzoek.

De Heer A. Brandenburg vervaardigde de tekeningen voor dit proefschrift.

Mijn bijzondere dank gaat tenslotte uit naar mijn ouders, voor het vele dat zij voor mij gedaan hebben. Aan hen zij dit proefschrift dan ook opgedragen.

CONTENTS

CHAPTER 1

INTRODUCTION AND SURVEY OF LITERATURE

INTRODUCTION

The present thesis originates from the interest in the Laboratory for Organic Chemistry of the University of Amsterdam in the chemistry of nitroso compounds. The results of the investigations in this field have been published in the series "C-Nitroso Compounds," starting in 1966.[1]

The synthesis of nitroso(cyclo)alkanes by photochemical nitrosation of the corresponding hydrocarbons with alkyl nitrites as nitrosating agents has been studied by Mackor[2] and kinetic and spectroscopic aspects of C-nitroso compounds have been investigated by Wajer.[3]

One of the most important properties of nitrosoalkanes is their ability to trap short-living radicals in solution to produce nitroxides. Mackor and Wajer[2,3] have studied several types of nitroxides R(X)NO· and there is a continuing interest in the properties of these species.

In current research nitroxides play an important role in Spin Labeling investigations in biological systems[4] and in Spin Trapping studies in organic chemistry.[5,6]

In Spin Labeling studies stable paramagnetic species are used to study *diamagnetic* systems, *e.g.* biological macromolecules. When for instance a stable (*i.e.* isolated) nitroxide which also contains a suitable functional group (like hydroxyl or amino) is allowed to react with the macromolecule,

---

[1] Papers in the series: "C- (and N-) Nitroso Compounds" have appeared in Tetrahedron, Tetrahedron Letters and Recueil des Travaux Chimiques des Pays-Bas. Part I has been published in 1966: A. Mackor, Th. A. J. W. Wajer, Th. J. de Boer and J. D. W. van Voorst, *Tetrahedron Letters* 2115 (1966). Recently Part XXII has appeared: J. A. Maassen, H. Hittenhausen and Th. J. de Boer, *Tetrahedron Letters* 3213 (1971).

[2] A. Mackor, Thesis, University of Amsterdam (1968).

[3] Th. A. J. W. Wajer, Thesis, University of Amsterdam (1969).

[4] O. H. Griffith and A. S. Waggoner, *Accounts Chem. Res.* 2, 17 (1969).

[5] E. G. Janzen, *Accounts Chem. Res.* 4, 31 (1971).

[6] M. J. Perkins in "Essays on Free-radical Chemistry," *Special Publ.* 24, The Chemical Society, London (1970), ch. 5.

information on the molecular structure of the latter may be deduced from the ESR spectrum of the paramagnetic adduct.

Spin Trapping is essentially different since here we use diamagnetic species to study *paramagnetic* systems, *e.g.* short-living free radicals. Generally speaking, Spin Trapping (which is the subject of the present thesis) is based on the effective radical scavenging properties of certain diamagnetic compounds. It turns out that monomeric nitrosoalkanes are highly efficient in this respect; information on the structure of the trapped radical may be deduced from the ESR spectrum of the paramagnetic adduct.

After a short description of the main features of nitrosoalkane and nitroxide chemistry an introduction will be given of the ESR Spin Trapping technique. Three studies in which this technique has been used, will be presented in chapters 2, 3 and 4. It will be shown that Spin Trapping is a valuable new tool for studying mechanisms of reactions which proceed *via* free radicals.

NITROSOALKANES

General properties

Crystalline nitroso(cyclo)alkanes usually exist in the colourless *trans*-dimeric form I. In solution this dimer ($\pi \rightarrow \pi^*$ absorption at 290–300 nm, $\log \varepsilon \cong 4.00$) is in equilibrium with the blue monomer II, which is characterized by a weak $n_N \rightarrow \pi^*$ absorption between 650 and 750 nm ($\varepsilon = 20$–50).

The dimer is dissociated to a large extent when a tertiary nitrosoalkane (R = *t*-alkyl) is dissolved in an organic solvent and these solutions are therefore coloured blue. Solutions of primary and secondary nitrosoalkanes are colourless and contain only traces of the monomer at room temperature. However the equilibrium may be shifted somewhat to the monomer side either thermally[7] or photochemically ($\lambda \cong 310$ nm).[8]

Primary and secondary nitrosoalkanes can undergo isomerization to the corresponding oximes, especially in the presence of acids or bases.

$$RR'CHNO \longrightarrow RR'C{=}NOH$$

[7] Th. A. J. W. Wajer, A. Mackor, Th. J. de Boer and J. D. W. van Voorst, *Tetrahedron* **23**, 4021 (1967).
[8] A. Mackor, Th. A. J. W. Wajer and Th. J. de Boer, *Tetrahedron Letters* 2757 (1967).

However, in the absence of catalysts the energy of activation is high and dimerization is favoured.

Literature

The synthesis, properties and reactions of nitrosoalkanes have been reviewed in a volume edited by Feuer[9] and, with the emphasis on synthetic aspects, by Metzger and Meier.[10]

NITROXIDES

Structure

Nitroxides are paramagnetic species containing the N–O moiety with one unpaired electron. For organic purposes the representation by a resonance hybrid of III and IV is most convenient, consistent with the nitrogen-oxygen bond order of approximately 1.5, as obtained from infrared spectral data.[11]

However, the structure may also be represented by V in which a three electron bond is involved or by VI in which the hypothesis of Linnett[12] is used. The particular electronic arrangement in these structures is inherent with the stability of nitroxides and their resistance to dimerization.

The term *nitroxide* is most often encountered in the literature and will be used throughout this thesis. However, some investigators favour an other nomenclature and the names *nitroxyl* and *iminoxyl* (the latter for non-aromatic nitroxyls) have also been adopted (mainly in soviet literature[13]) in order to emphasize the radical nature of these species.

---

[9] H. Feuer, "The Chemistry of Nitro and Nitroso Groups," Part I, Interscience, New York (1969).

[10] H. Metzger and H. Meier in Houben-Weyl, "Methoden der Organischen Chemie," Band X/1, Thieme Verlag, Stuttgart (1971), p. 891.

[11] A. R. Forrester, J. M. Hay and R. H. Thomson, "Organic Chemistry of Stable Free Radicals," Academic Press, London (1968), ch. 5.

[12] For details of the so called "double quartet" representation, cf. J. W. Linnett, *J. Amer. Chem. Soc.* 83, 2643 (1961); J. W. Linnett and R. M. Rosenberg, *Tetrahedron* 20, 53 (1964).

[13] E. G. Rozantsev and V. D. Sholle, *Synthesis* 190 (1971).

Hyperfine splitting due to the nitrogen nucleus

Since ^{14}N has I = 1, the ESR spectra of nitroxides (and other nitrogen-centered radicals) are always characterized either by three equidistant lines of the same intensity, or by three groups of lines when the unpaired electron also interacts with other nuclei possessing a nuclear spin [*e.g.* hydrogen (I = $\frac{1}{2}$), fluorine (I = $\frac{1}{2}$), chlorine (I = $\frac{3}{2}$) and bromine (I = $\frac{3}{2}$)]. This causes the so-called hyperfine splitting (hfs) which plays an essential role in the spectra of the nitroxides throughout the present thesis.

The contributions of structure III and IV are dependent on the polarity and the hydrogen bonding ability of the solvent; the magnitude of the nitrogen hfs constant a_N is slightly larger in the more polar media where structure III is favoured with the unpaired electron located on the nitrogen atom. The higher spin density on nitrogen results in a larger a_N value.

The values of the nitrogen hfs constants are influenced by conjugative and inductive effects of R and R'; the a_N value therefore has considerable diagnostic value. Whereas di-alkyl nitroxides show a characteristic a_N value of about 15 gauss, aryl alkyl nitroxides have a_N values of about 12.5 gauss and di-aryl nitroxides have a_N values around 10 gauss.[7,11,14] Both unpaired electron and electron pair delocalization would explain the decreasing a_N values, although the latter is presumably favoured.[15]

Acyl alkyl nitroxides have rather low a_N values of about 7.5 gauss.[16] This indicates a smaller spin density on the nitrogen atom and consequently a larger spin density on oxygen, since resonance structure IX is expected to participate in π-electron (pair) delocalization.

In this representation we assume that delocalization of an electron pair onto the carbonyl moiety is again more favourable than unpaired electron delocalization.

[14] Th. A. J. W. Wajer, H. W. Geluk, J. B. F. N. Engberts and Th. J. de Boer, *Rec. Trav. Chim.* **89**, 696 (1970); G. Rawson and J. B. F. N. Engberts, *Tetrahedron* **26**, 5653 (1970).

[15] R. I. Walter, *J. Amer. Chem. Soc.* **88**, 1923 (1966).

[16] A. Mackor, Th. A. J. W. Wajer and Th. J. de Boer, *Tetrahedron* **24**, 1623 (1968).

Di-acyl nitroxides may have a_N values as low as 4 gauss; *e.g.* the nitroxide obtained by oxidation of *N*-hydroxyphthalimide has $a_N = 4.2$ gauss.[17]

In some other classes of nitroxides the configuration at nitrogen may not be planar, resulting in larger a_N values due to the increased *s*-character of the orbital which contains the unpaired electron. The a_N values of sulfonyl nitroxides have been discussed in these terms.[14]

g-Values

The *g*-value of a free radical is largely determined by spin-orbit coupling and this effect increases when the unpaired electron interacts with atoms of increasing atomic number (*i.e.* oxygen > nitrogen).[13] The *g*-value therefore has some diagnostic value; *e.g.* acyl alkyl nitroxides are not only characterized by their low a_N values, but also by a *g*-value of about 2.0067 which is larger than the one for di-alkyl nitroxides ($g = 2.0060$), since the unpaired electron is located to a larger extent on the NO oxygen atom (and possibly on the CO oxygen atom) in acyl alkyl nitroxides.

Nomenclature

The following nomenclature is used to indicate the position of substituents, as exemplified for hydrogen atoms.

$$-\overset{|}{\underset{H\gamma}{C}} - \overset{|}{\underset{H\beta}{C}} - \overset{•}{\underset{H\alpha}{N}} - O$$

Generation of nitroxides

In solution *C*-nitroso compounds X readily form di-alkyl nitroxides XI upon photolysis or heating.[7] This reaction proceeds particularly well with tertiary nitrosoalkanes, showing that the monomeric form is involved. Apparently homolytic scission of the carbon-nitrogen bond takes place and the alkyl radical is trapped by a second monomer molecule to produce the nitroxide.[18]

$$R-N{=}O \xrightarrow[\text{or } \Delta]{h\nu} R^{\bullet} + NO \xrightarrow{R-N=O} R-\underset{\underset{O}{|\bullet}}{N}-R$$

$$\text{X} \qquad\qquad\qquad\qquad\qquad \text{XI}$$

[17] H. Lemaire and A. Rassat, *J. Chim. Phys.* **61**, 1580 (1964).
[18] J. A. Maassen, H. Hittenhausen and Th. J. de Boer, *Tetrahedron Letters* 3213 (1971).

The di-alkyl nitroxides are not only characterized by their a_N value of 14–16 gauss, but very often other splittings due to hydrogen atoms in the alkyl groups simplify their identification. β-Hydrogen atoms always show hfs in the spectra, the magnitude being strongly dependent upon the conformation; *i.e.* the magnitude of $a_{H\beta}$ is related to the dihedral angle, θ, between the planes defined by the axis of the nitrogen $2p_z$-orbital, the N–CH$_\beta$ and the C–H$_\beta$ bond (*cf.* chapter 2).[19]

Generally no hfs due to γ-hydrogen atoms is resolved, *e.g.* the spectrum of di-*t*-butyl nitroxide consists of only the three nitrogen lines when measured in solution at room temperature.[20] However, examples are known in which hfs due to γ- and δ-hydrogen atoms is resolved; the configuration is very important in these long-range couplings. The coupling with δ-hydrogen atoms is only favourable when the system is in a W-shape configuration.[19,21]

Literature

The synthesis and the chemistry of stable nitroxides have recently been reviewed by Forrester *et al.*[11] and by Rozantsev.[13,22]

ESR SPIN TRAPPING TECHNIQUES

Introduction

Short-lived free radicals may be studied directly by ESR spectroscopy when a sufficiently high concentration is produced either by using a rapid-mixing flow system[23] or by intense UV irradiation in a static system at very low temperatures.[24] These methods have not been readily adopted in organic chemistry, since they have limited applicability and are often laborious.

Since 1966 several papers of Mackor and Wajer *et al.*[7,16,25] have appeared

[19] Th. A. J. W. Wajer, Thesis, University of Amsterdam (1969), p. 123; Th. A. J. W. Wajer, A. Mackor and Th. J. de Boer, *Rec. Trav. Chim.* **90**, 568 (1971).

[20] At temperatures below −50° the γ-hydrogen hfs has been partially resolved; *cf.* Th. A. J. W. Wajer, Thesis, University of Amsterdam (1969), p. 130–132.

[21] C. Morat and A. Rassat, *Bull. Soc. Chim. France* 893 (1971).

[22] E. G. Rozantsev, "Free Nitroxyl Radicals," Plenum Press, New York (1970).

[23] W. T. Dixon and R. O. C. Norman, *J. Chem. Soc.* 3119 (1963); R. O. C. Norman in "Essays on Free-radical Chemistry," *Special Publ.* 24, The Chemical Society, London (1970), ch. 6.

[24] J. K. Kochi, P. J. Krusic and D. R. Eaton, *J. Amer. Chem. Soc.* **91**, 1877 (1969); J. K. Kochi in "Essays on Free-radical Chemistry," *Special Publ.* 24, The Chemical Society, London (1970), ch. 7.

[25] A. Mackor, Th. A. J. W. Wajer, Th. J. de Boer and J. D. W. van Voorst, *Tetrahedron Letters* 2115 (1966); *ibid.* 385 (1967).

in which the three types of paramagnetic intermediates are described, which are formed during the photochemical nitrosation of hydrocarbons by alkyl nitrites. Though these results will be described more in detail in chapter 3, it is relevant at this point to mention that these nitroxides were formed by addition of free radicals (alkyl, alkoxy and acyl) to nitrosoalkanes (prepared *in situ*). In fact these results show the excellent scavenging properties of monomeric nitrosoalkanes and in the last two or three years several groups of (organic) chemists[26-32] have started investigations in the field of ESR Spin Trapping. This technique is based on the idea of deliberately adding a small amount of a diamagnetic scavenger to a reacting system in order to trap highly reactive, *i.e.* short-living radicals. The method is only useful when the spin adduct is more stable than the trapped radical itself.

Since nitroxides belong to the more stable types of free radicals, especially those scavengers have been used which ultimately produce such nitroxides (*cf.* following section). In most cases the spectra of these nitroxides may be conveniently measured at room temperature in a static system.

Since Spin Trapping is experimentally simple and less time-consuming than the direct observation of short-lived free radicals (using flow-systems *etc.*), it is a powerful tool for studying reaction mechanisms in which free radicals are involved.

Recently an other sensitive method for studying short-living free radicals has been developed, *i.e.* chemically induced dynamic nuclear polarization (CIDNP).[33]

Types of scavengers used for Spin Trapping

Two types of scavengers have been used frequently:[34] nitrosoalkanes XII[26-29] and nitrones XIV.[30,31]

---

[26] C. Lagercrantz and S. Forshult, *Nature* **218**, 1247 (1968).

[27] G. R. Chalfont, M. J. Perkins and A. Horsfield, *J. Amer. Chem. Soc.* **90**, 7141 (1968).

[28] I. H. Leaver and G. C. Ramsay, *Tetrahedron* **25**, 5669 (1969).

[29] J. W. Hartgerink, J. B. F. N. Engberts, Th. A. J. W. Wajer and Th. J. de Boer, *Rec. Trav. Chim.* **88**, 481 (1969).

[30] M. Iwamura and N. Inamoto, *Bull. Chem. Soc. Japan* **40**, 702 (1967).

[31] E. G. Janzen and B. J. Blackburn, *J. Amer. Chem. Soc.* **90**, 5909 (1968).

[32] H. J. Jakobsen and K. Torssell, *Tetrahedron Letters* 5003 (1970).

[33] R. Kaptein, Thesis, University of Leiden (1971).

[34] Other scavengers have occasionally been applied, *inter alia* nitric oxide and the *aci*-anion of nitromethane [*cf.* N. H. Anderson and R. O. C. Norman, *J. Chem. Soc.* (*B*) 993 (1971)] and aromatic nitrile *N*-oxides [*cf.* B. C. Gilbert, V. Malatesta and R. O. C. Norman, *J. Amer. Chem. Soc.* **93**, 3290 (1971)].

$$R-N=O \quad + \quad X^{\bullet} \longrightarrow R-\underset{|\bullet}{\overset{}{N}}-X$$

$$\underline{\text{XII}} \qquad\qquad\qquad \underline{\text{XIII}} \quad \overset{O}{}$$

$$R-\underset{\overset{|}{\underset{\ominus}{\text{O}}}}{\overset{\oplus}{N}}=C\underset{R''}{\overset{R'}{\diagup}} \quad + \quad X^{\bullet} \longrightarrow R-\underset{\overset{|\bullet}{O}}{\overset{}{N}}-\underset{\overset{|}{X}}{\overset{R'}{C}}-R''$$

$$\underline{\text{XIV}} \qquad\qquad\qquad\qquad \underline{\text{XV}}$$

Both produce nitroxides; the trapped radical X is attached directly to the nitrogen atom in XIII, whereas it is linked to the α-carbon atom when a nitrone is used as a scavenger. Generally this means that elucidation of the structure of the trapped radical is easier when a nitrosoalkane has been used as a scavenger, since X is nearer to the NO moiety containing the unpaired electron and will show resolved hfs more readily.

However, one should not conclude that a nitrosoalkane is always the best choice in Spin Trapping, as exemplified by the fact that certain radicals (*e.g.* halogen atoms) are trapped by nitrones and *not* by nitrosoalkanes.[5]

Nitrosoalkanes as scavengers

Since the scavenging of radicals by nitrosoalkanes is a property of the monomeric form, it is clear that *tertiary* nitrosoalkanes, which monomerize in solution, have been used by most investigators.[26-29] An other advantage of tertiary nitrosoalkanes is the fact that no isomerization to oximes is possible. In addition, the interpretation of the ESR spectra of the derived nitroxides XIII is simpler since no β-hydrogen atoms are present in the alkyl moiety (once incorporated in the nitroxide).

For these reasons 2-nitroso-2-methylpropane (XII, R = *t*-butyl; "*t*-nitrosobutane") has been used in many studies, including the present thesis. When hfs is resolved in the spectrum (apart from the nitrogen splitting) it is always caused by the group X, since the nine equivalent γ-hydrogen atoms of the *t*-butyl moiety do only cause some line-broadening. Very recently perdeutero-2-nitroso-2-methylpropane has also been used a as scavenger[32,35] and some small splittings which were obscured by line-broadening in the undeuterated species could be resolved. An example is discussed in chapter 3.

Nitrones as scavengers

Nitrones have been used as scavengers for short-living radicals by Iwamura and Inamoto[30] and, subsequently, by the group of Janzen.[5,31] Both α,*N*-

[35] R. J. Holman and M. J. Perkins, *Chem. Commun.* 244 (1971); *idem, J. Chem. Soc. (C)* 2324 (1971).

8

diphenylnitrone (XIV, R = R′ = C_6H_5, R″ = H) and α-phenyl-*N-t*-butyl-nitrone (XIV, R = *t*-C_4H_9, R′ = C_6H_5, R″ = H) have been employed.[30,36] The *N*-phenyl group in the first nitrone complicates analysis of the spectra of the derived nitroxides and therefore the second nitrone with an *N-t*-butyl group is preferred.

Literature

Recently two reviews concerning Spin Trapping have appeared. The review by Janzen[5] is the more general one; the advantages and disadvantages of the use of nitrosoalkanes *versus* nitrones are discussed with an emphasis on nitrones. In the review by Perkins[6] the emphasis is on the use of nitroso-alkanes as scavengers.

---

[36] Recently *N-t*-butylnitrone (XIV, R = *t*-C_4H_9, R′ = R″ = H) has also been used in Spin Trapping studies; *cf.* G. R. Chalfont, M. J. Perkins and A. Horsfield, *J. Chem. Soc.* (*B*) 401 (1970); also ref. 34.

TRAPPING OF TRIFLUOROMETHYL RADICALS BY MONOMERIC NITROSOALKANES

SYNTHESIS OF 1-NITROSOADAMANTANE

Abstract – Photolysis of trifluoroiodomethane in benzene produces trifluoromethyl radicals, which can be trapped by monomeric nitrosoalkanes (RNO) to produce alkyl trifluoromethyl nitroxides [R(CF$_3$)NO$^{\cdot}$]. These nitroxides are very stable, in sharp contrast with analogous nitroxides which contain halogen atoms other than fluorine.[2] The ESR spectra of these nitroxides show rather large fluorine hyperfine splittings (a_F = 11.5–12.0 gauss). The possible mechanisms for spin delocalization oy β-fluorine atoms are discussed. One of the scavengers used in this study is hitherto unknown 1-nitrosoadamantane.

INTRODUCTION

Some years ago Blackley and Reinhard[3] found that di-trifluoromethyl nitroxide (prepared by oxidation of the corresponding hydroxylamine) is a remarkably stable radical. This was ascribed to the strong electronegative character of the trifluoromethyl groups and the possible delocalization of the unpaired electron onto the six fluorine atoms.

The interaction between a trifluoromethyl group and a neighbouring π-electron system (*i.e.* a phenyl group) has been studied by Sheppard[4] who first suggested through-space interaction between a fluorine 2p-orbital and the π-electron system.

Scheidler and Bolton[5] showed that the nitrogen and fluorine hfs constants of di-trifluoromethyl nitroxide change in opposite directions when the temperature is lowered from $+4°$ to $-110°$. The increase of the fluorine hfs constants (from 8.26 to 8.76 gauss) was explained by more efficient fluorine-nitrogen p–π interaction, because at low temperatures the configu-

---

[1] Part of this chapter has been published in a preliminary communication: J. W. Hartgerink, J. B. F. N. Engberts, Th. A. J. W. Wajer and Th. J. de Boer, *Rec. Trav. Chim.* **88**, 481 (1969).

[2] The latter type of nitroxides will be the subject of chapter 3.

[3] W. D. Blackley and R. R. Reinhard, *J. Amer. Chem. Soc.* **87**, 802 (1965).

[4] W. A. Sheppard, *J. Amer. Chem. Soc.* **87**, 2410 (1965).

[5] P. J. Scheidler and J. R. Bolton, *J. Amer. Chem. Soc.* **88**, 371 (1966).

rations favourable for such interactions were supposed to be relatively more populated. In this case p–π interaction may take place between a fluorine $2p$-orbital and the nitrogen $2p_z$-orbital which is part of the π-system containing the unpaired electron.

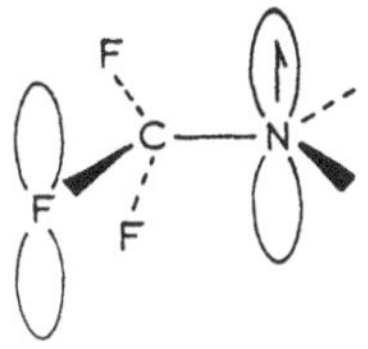

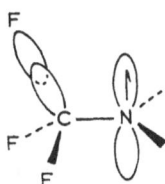

Figure 1. The fluorine-nitrogen p–π overlap mechanism.

It is interesting to note that this interaction is most efficient in the conformation depicted in figure 1, the dihedral angle, θ, between the two planes defined by the axis of the nitrogen $2p_z$-orbital, the N–CF$_\beta$ and the C–F$_\beta$ bond being 90°.

A completely different dependence on the dihedral angle is expected in the alternative mechanism for the interaction of the unpaired electron with β-fluorine atoms. In the hyperconjugation mechanism there is interaction between the nitrogen $2p_z$-orbital containing the unpaired electron and the carbon-fluorine σ-bond.

Figure 2. The hyperconjugation mechanism.

This mechanism is analogous to that proposed for the interaction of β-hydrogen atoms with the nitrogen $2p_z$-orbital in di-alkyl nitroxides;[6,7] it has been shown that the $a_{H\beta}$ in di-alkyl nitroxides depends on the dihedral angle θ (Heller-McConnell relationship[8]), $a_{H\beta}$ being proportional to the time-averaged value of $cos^2\,\theta$ (*i.e.* $\langle cos^2\,\theta \rangle$). The most favourable confor-

[6] G. Chapelet-Letourneux, H. Lemaire, R. Lenk, M. A. Maréchal and A. Rassat, *Bull. Soc. Chim. France* 3963 (1968).

[7] Th. A. J. W. Wajer, Thesis, University of Amsterdam (1969), p. 123.

[8] C. Heller and H. M. McConnell, *J. Chem. Phys.* **32**, 1535 (1960).

11

mation for hyperconjugation is the one with $\theta = 0°$, analogous to the one depicted in figure 2.

The magnitude of the $a_{H\beta}$ values in the following series of nitroxides has been shown to decrease dramatically when the degree of substitution at the α-carbon atom gradually increases.[7]

$$
\begin{array}{ccc}
\text{H} & \text{CH}_3 & \text{CH}_3 \\
| & | & | \\
\text{H—C—N—t-Bu} & \text{H—C—N—t-Bu} & \text{CH}_3\text{—C—N—t-Bu} \\
|\quad |• & |\quad |• & |\quad |• \\
\text{H}\quad\text{O} & \text{H}\quad\text{O} & \text{H}\quad\text{O}
\end{array}
$$

$a_{H\beta} = 11.7$ gauss (3H) $a_{H\beta} = 9.5$ gauss (2H) $a_{H\beta} = 1.5$ gauss (1H)

It appears that those conformations which are favourable for hyperconjugation become less populated upon higher substitution, resulting in smaller $\langle cos^2\ \theta \rangle$ values and therefore smaller $a_{H\beta}$ values.

Strom and Bluhm[9] have compared the fluorine hfs constants of I and II and found the highest $a_{F\beta}$ value for compound II with the more bulky substituents.

$$
\begin{array}{cc}
\text{F}\qquad\text{F} & \text{R}\qquad\text{R} \\
|\qquad| & |\qquad| \\
\text{F—C—N—C–F} & \text{F—C—N—C–F} \\
|\ |•\ | & |\ |•\ | \\
\text{F}\ \text{O}\ \text{F} & \text{F}\ \text{O}\ \text{F} \\
\text{I} & \text{II}
\end{array}
$$

$a_{F\beta} = 8.2$ gauss (6F)[3]
R = CF$_2$CO$_2$CH$_3$,
$a_{F\beta} = 13.8$ gauss (4F)[9]

Because the substituent effect is *opposite* to that observed in the case of β-hydrogen atoms, they favoured the p–π interaction mechanism.

Underwood et al.[10] extended their study to higher substituted compounds (*vide infra*, III to V) and suggested that hyperconjugation can equally well explain the substituent effect on $a_{F\beta}$ by assuming a different substituent effect on conformation-populations in fluorine compounds on the one hand and on non-fluorine compounds on the other.

$$
\begin{array}{ccc}
\text{F} & \text{R} & \text{R} \\
| & | & | \\
\text{F—C—N—t-Bu} & \text{F—C—N—t-Bu} & \text{R—C—N—t-Bu} \\
|\ |• & |\ |• & |\ |• \\
\text{F}\ \text{O} & \text{F}\ \text{O} & \text{F}\ \text{O} \\
\text{III} & \text{IV} & \text{V}
\end{array}
$$

$a_{F\beta} = 11.8$ gauss (3F)[10]
R = CF$_3$,
$a_{F\beta} = 21.9$ gauss (2F)[11]
R = CF$_3$,
$a_{F\beta} = 2.2$ gauss (1F)[10]

[9] E. T. Strom and A. L. Bluhm, *Chem. Commun.* 115 (1966).
[10] G. R. Underwood, V. L. Vogel and I. Krefting, *J. Amer. Chem. Soc.* **92**, 5019 (1970).
[11] K. J. Klabunde, *J. Amer. Chem. Soc.* **92**, 2427 (1970).

Definite conclusions on the mechanism for the fluorine-nitrogen interaction can apparently not be drawn at this moment; it may well be that
both mechanisms are operative, the relative importance of either depending
on the nature of the compound.

As part of the interest in β-halogenated nitroxides[12] we studied the
photolysis of trifluoroiodomethane in the presence of nitroso(cyclo)alkanes
in order to obtain alkyl trifluoromethyl nitroxides in a simple way. In the
course of our work this method (*i.e.* photolysis of perfluoroalkyl iodides in
the presence of nitrosoalkanes) has also been applied by other investigators.[10,11]

RESULTS AND DISCUSSION

Trifluoroiodomethane is known to be a clean source of trifluoromethyl
radicals.[13] We found that they can be trapped efficiently by monomeric
nitrosoalkanes VII to produce alkyl trifluoromethyl nitroxides VIII. The
iodine atoms which are also produced are not trapped at all (*or* nitroxide
IX decomposes too quickly to be detectable); this is in agreement with the
observations of Janzen who reported that, despite several attempts, so far
no halogen atoms have been trapped by nitrosoalkanes.[14]

The ultraviolet irradiation used for the photolysis of trifluoroiodomethane can also dissociate nitroso-dimers VI (*e.g.* R = *sec.* alkyl) into the
reactive monomers VII and secondary nitrosoalkanes may therefore also
be used as scavengers.

The hfs constants of several alkyl trifluoromethyl nitroxides derived from
secondary and tertiary nitrosoalkanes are collected in table 1.

In all cases the nitrogen and the fluorine hfs constants are practically
equal, resulting in an intensity distribution which is in very close agreement
with the theoretically expected 1:4:7:7:4:1 intensity ratios. The alkyl
trifluoromethyl nitroxides are rather stable species (*cf.* Experimental).

[12] Th. A. J. W. Wajer, Thesis, University of Amsterdam (1969), p. 95.
[13] R. N. Haszeldine, *J. Chem. Soc.* 584 (1951).
[14] E. G. Janzen, *Accounts Chem. Res.* 4, 31 (1971).

13

TABLE 1

HFS constants (in gauss) of alkyl trifluoromethyl nitroxides [R(CF$_3$)NO·] formed in benzene at 20° by photolysis of trifluoroiodomethane in the presence of various nitrosoalkanes

	R	a_N	$a_{H\beta}$	a_F
tert.	R = 1-adamantyl[a]	12.0	–	12.0 (3F)
	t-butyl[b]	11.8	–	11.8 (3F)
	2-(2,3-dimethyl)butyl	11.7	–	11.7 (3F)
sec.	R = *i*-propyl	11.5	2.8 (1H)	11.5 (3F)
	cyclopentyl	11.5	3.3 (1H)	11.5 (3F)
	cyclohexyl	11.6	3.0 (1H)	11.6 (3F)

[a] $g = 2.0063$.

[b] Ref. 11 reports $a_N = a_F = 12.05$ gauss (in benzene) and ref. 10 $a_N = a_F = 11.8$ gauss (solvent not reported).

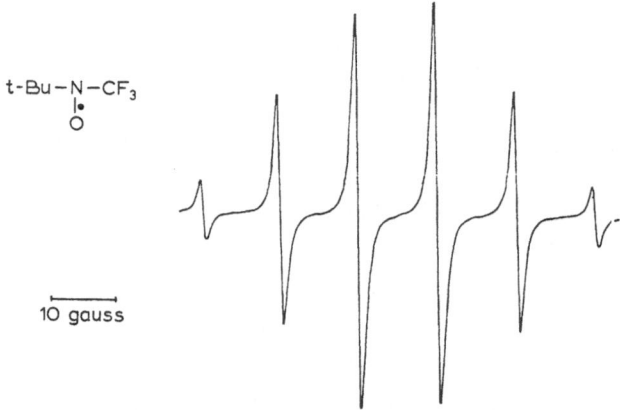

t-Bu–N–CF$_3$
|•
O

10 gauss

Figure 3. ESR spectrum of *t*-butyl trifluoromethyl nitroxide formed by photolysis of CF$_3$I in the presence of 2-nitroso-2-methylpropane in benzene at 20°.

No significant changes are observed when the spectra are measured over a range of temperatures. Klabunde[11] studied the temperature dependence (between +40° and −90°) of the spectrum of *t*-butyl trifluoromethyl nitroxide (VIII, R = *t*-butyl) and he also observed only minor changes in the hfs constants.

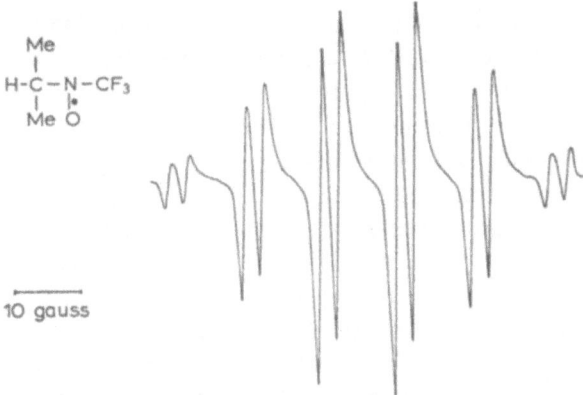

Figure 4. ESR spectrum of *i*-propyl trifluoromethyl nitroxide formed by photolysis of CF$_3$I in the presence of 2-nitrosopropane in benzene at 20°.

1-*Nitrosoadamantane*

One of the nitrosoalkanes used as a spin trapping agent (*cf.* table 1) is the previously unreported 1-nitrosoadamantane X. This compound can be prepared from 1-aminoadamantane using modifications of known synthetic methods.[15] Oxidation of 1-aminoadamantane with potassium permanganate in aqueous acetone affords 1-nitroadamantane.[16] Subsequent reduction with zinc and ammonium chloride gives *N*-1-adamantyl hydroxylamine (not isolated), which is oxidized by potassium dichromate–sulfuric acid to X, isolated by steam distillation as the colourless *trans*-dimer. Compound X has also been isolated by Morat and Rassat[17] as a byproduct in the synthesis of 1-adamantyl *t*-butyl nitroxide from 1-nitroadamantane and *t*-butyl magnesium chloride. However, their nitrosocompound may have been contaminated with an impurity as is concluded from our infrared spectral data.

For nitrosoadamantane, prepared *via* the hydroxylamine route, we find the following absorptions (in chloroform): 1250 cm^{-1} (NO, dimer) and

[15] J. R. Schwartz, *J. Amer. Chem. Soc.* **79**, 4353 (1957).
[16] H. Stetter, J. Mayer, M. Schwarz and K. Wulff, *Chem. Ber.* **93**, 226 (1960).
[17] C. Morat and A. Rassat, *Bull. Soc. Chim. France* 891 (1971).

1545 cm^{-1} (NO, monomer); no absorptions between 1545 and 2800 cm^{-1}. In a KBr-disc the peak at 1545 cm^{-1} is hardly present, whereas the absorption at 1250 cm^{-1} is much stronger. This is in agreement with the fact that only dimer is present in the solid-state, whereas an equilibrium between dimer and monomer exists in solution.

The product of Morat and Rassat[17] (no microanalysis reported) shows in nujol *inter alia* infrared absorptions (tentative assignments between parentheses) at 1250 cm^{-1} (NO, dimer), 1530 cm^{-1} (impurity) and 1620 cm^{-1} (NO, monomer). Our experience is that the pure dimer (microanalysis within 0.16%) does not produce an absorption at 1620 cm^{-1} upon dissolution in chloroform and so the absorption found by Morat and Rassat should be attributed to an impurity. Apparently the 1530 cm^{-1} absorption reported by these investigators corresponds to our 1545 cm^{-1} absorption (NO, monomer).

In chapter 1 we have mentioned the easy formation of di-alkyl nitroxides by irradiation of especially tertiary nitrosoalkanes. The fact that these species are not observed during photolysis of trifluoroiodomethane in the presence of (tertiary) nitrosoalkanes illustrates the efficiency of C-nitroso compounds for trapping trifluoromethyl radicals. Klabunde[11] estimates that this efficiency is at least 80%; this statement is based on his observation that irradiation of a sample which contains just a slight excess of a perfluoroalkyl iodide over the amount of 2-nitroso-2-methylpropane *only* produces the perfluoroalkyl *t*-butyl nitroxide signal.

Photolysis of X in benzene at 20° in the absence of external radical sources very easily affords di-1-adamantyl nitroxide XI (figure 5).[18] The a_N value of 14.5 gauss[19] is in accordance with that expected for a di-alkyl nitroxide. When di-1-adamantyl nitroxide is generated in low concentration, the ESR spectrum also shows resolved hfs due to twelve equivalent adamantyl γ-hydrogen atoms ($a_H = 0.43$ gauss; *cf.* figure 6).

The synthesis and isolation of XI by condensation of 1-nitroadamantane and 1-bromoadamantane in the presence of sodium have been described by Morat and Rassat,[17,20] who reported the ESR as well as the NMR spectral data in several solvents. In carefully deoxygenated ethanol the ESR spectrum showed some additional hfs due to six equivalent δ-hydrogen atoms). It was only possible to interpret this more complex spectrum with the help

---

[18] The stability of the 1-adamantyl radical which is trapped in this reaction has been assessed and is found to be comparable to the stability of the *t*-butyl radical; *cf.* J. P. Lorand, S. D. Chodroff and R. W. Wallace, *J. Amer. Chem. Soc.* **90**, 5266 (1968).

[19] Ref. 17 reports $a_N = 15.2$ gauss (in benzene).

[20] C. Morat and A. Rassat, *Bull. Soc. Chim. France* 893 (1971).

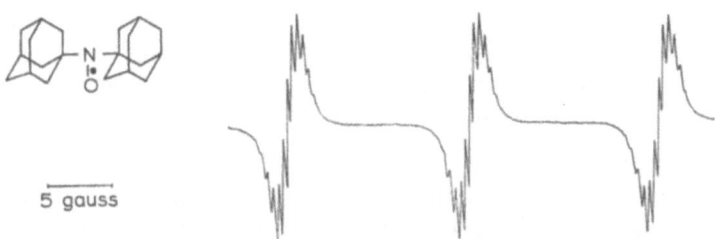

Figure 5. ESR spectrum of di-1-adamantyl nitroxide formed by photolysis of 1-nitrosoadamantane in benzene at 20°.

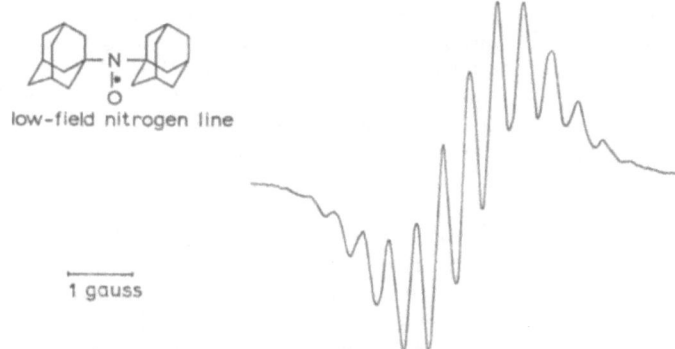

Figure 6. Low-field nitrogen line of the spectrum of di-1-adamantyl nitroxide.

of computer simulation.[20] The best fit with the experimental spectrum (in ethanol) was obtained with the following set of parameters: a_H = 0.41 gauss (12 γ–H), a_H = 0.57 gauss (6 δ–H) and a linewidth of 0.23 gauss.

EXPERIMENTAL

All melting points and boiling points are uncorrected. IR spectra were recorded with a Unicam SP 200 spectrometer. NMR spectra were determined on a Varian A-60 instrument with TMS as an internal standard ($\delta = 0$) and $CDCl_3$ or CCl_4 as the solvent. ESR spectra were measured on a Varian E-3 apparatus fitted with an optical transmission cavity; the light source was a Philips SP 500 W super high pressure mercury lamp.

Materials

Trifluoroiodomethane was prepared by Hunsdiecker decarboxylation of the silver salt of trifluoroacetic acid in the presence of iodine[13] and was isolated in a cold trap as a pale purple liquid, b.p. −22° (lit.[13] b.p. −22°). A dilute solution in benzene was used for the spin trapping experiments.

17

C-nitroso compounds

The following C-nitroso compounds were prepared by photochemical nitrosation of the corresponding hydrocarbons with t-butyl nitrite and were isolated as the trans-dimers: nitrosocyclopentane,[21] nitrosocyclohexane[22] and 2-nitroso-2,3-dimethylbutane.[23,24] 2-Nitrosopropane[25] was prepared via the oxaziridine according to Emmons.[26] 2-Nitroso-2-methylpropane ("t-nitrosobutane") was prepared by reduction of 2-nitro-2-methylpropane with zinc and ammonium chloride to give N-2-(2-methyl)propyl hydroxylamine (not isolated); subsequent oxidation with potassium dichromate–sulfuric acid under simultaneous steam distillation produced 2-nitroso-2-methylpropane in an overall yield of 50%.

1-Nitrosoadamantane

1-Aminoadamantane was synthesized from adamantane[27] in three steps according to Stetter et al.[16,28] Oxidation of 1-aminoadamantane with potassium permanganate in aqueous acetone afforded 1-nitroadamantane[16] in 26% yield.

A mixture of 2.0 g of 1-nitroadamantane, 2.3 g of ammonium chloride, 2.2 g of zinc dust and 15 ml of water was vigorously shaken for 30 minutes; the temperature was kept below 50°. The mixture was filtered with suction; the solid residue was washed with several 10 ml portions of boiling water. The filtrate in which the N-1-adamantyl hydroxylamine started to crystallize was used for the next step without isolation of the hydroxylamine derivative. The filtrate was acidified with 0.7 ml of 2N sulfuric acid and heated to 100° and subsequently a solution of 1.1 g of potassium dichromate in 0.9 ml of conc. sulfuric acid and 3.7 ml of water was added dropwise under simultaneous steam distillation. Blue, monomeric 1-nitrosoadamantane was removed from the reaction flask and dimerized in the condenser to the colourless trans-dimer (trans-1-azodioxyadamantane), yield 0.46 g (25%). Infrared spectra: KBr-disc, 1250 cm^{-1} (NO, dimer); CHCl$_3$-solution, 1545 cm^{-1} (NO, monomer) and 1250 cm^{-1} (NO, dimer). NMR spectrum in accordance with that expected for a 1-substituted adamantane. Anal. Found: C, 72.79; H, 9.28; N, 8.32; O, 9.82; C$_{20}$H$_{30}$N$_2$O$_2$ (dimer) requires: C, 72.69; H, 9.15; N, 8.48; O, 9.68%.

ESR Spin Trapping experiments

A few crystals of a nitrosoalkane were dissolved in the dilute solution of trifluoroiodomethane in benzene (about 5 mg of nitrosoalkane per ml of solvent). After deoxygenation by a stream of nitrogen the sample tube was irradiated with ultraviolet light in the cavity of the ESR apparatus for a short time (about 30 sec.). The alkyl trifluoromethyl nitroxides proved to be rather stable. At room temperature the ESR signals could be easily maintained for many hours.

[21] A. Mackor and Th. J. de Boer, Rec. Trav. Chim. **89**, 151 (1970).

[22] A. Mackor, J. U. Veenland and Th. J. de Boer, Rec. Trav. Chim. **88**, 1249 (1969).

[23] A. Mackor and Th. J. de Boer, Rec. Trav. Chim. **89**, 159 (1970).

[24] Kindly provided by Dr. A. Mackor.

[25] Kindly provided by Dr. Th. A. J. W. Wajer.

[26] W. D. Emmons, J. Amer. Chem. Soc. **79**, 6522 (1957).

[27] I am indebted to N.V. Philips-Duphar Research Laboratories, Weesp, The Netherlands for a generous gift of adamantane.

[28] H. Stetter, M. Schwarz and A. Hirschhorn, Chem. Ber. **92**, 1629 (1959).

CHAPTER 3[1]

PHOTOLYSIS OF POLYHALOMETHANES IN THE PRESENCE OF NITROSOALKANES

FORMATION OF ACYL ALKYL NITROXIDES

Abstract – When tri-, di- and monohalomethyl radicals are generated photo-chemically in the presence of 2-nitroso-2-methylpropane the expected ESR spectra of the (poly)halomethyl t-butyl nitroxides are normally only observed when the halogen atom is fluorine (see chapter 2) and not when it is chlorine, bromine or iodine. These nitroxides are often very unstable. Apparently, β-halogen elimination occurs easily (except for fluorine) to give halonitrones. These halonitrones may act as precursors for the formation of new types of acyl alkyl nitroxides, which are actually observed ($a_N = 6.5$–7.0 gauss). A mechanism is proposed in which the halonitrones undergo light-induced isomerization to halo-oxaziridines followed by a bimolecular oxidation-reduction reaction leading to the observed acyl alkyl nitroxides.

INTRODUCTION

The paramagnetic intermediates, which are formed during the photo-chemical nitrosation of hydrocarbons by alkyl nitrites have been studied extensively.[2] Three types of radicals have been observed: di-alkyl nitroxides[3] ($a_N = 14.6$–15.3 gauss), alkoxy alkyl nitroxides[3] ($a_N = 26.9$–30.2 gauss) and acyl alkyl nitroxides[4] ($a_N = 7.4$–8.3 gauss). These radicals result from the addition of alkyl, alkoxy and acyl radicals to (monomeric) nitroso-alkanes.

The mode of formation of alkyl and alkoxy radicals is obvious; the acyl radicals are formed only when primary and secondary nitrite esters RONO are used. Primary alkoxy radicals $RCH_2O\cdot$ may produce aldehydes by β-C–H fission, whereas secondary alkoxy radicals $R_2C(H)O\cdot$ can give aldehydes by β-C–R fission.[4] Hydrogen abstraction from the aldehydes produces the acyl radicals. Since tertiary alkoxy radicals do not undergo these reactions, no acyl alkyl nitroxides have been observed during the

---

[1] Part of this chapter has been published as a short communication: J. W. Hartgerink, J. B. F. N. Engberts and Th. J. de Boer, *Tetrahedron Letters* 2709 (1971).

[2] A. Mackor, Thesis, University of Amsterdam (1968), p. 63–94.

[3] A. Mackor, Th. A. J. W. Wajer, Th. J. de Boer and J. D. W. van Voorst, *Tetrahedron Letters* 385 (1967).

[4] A. Mackor, Th. A. J. W. Wajer and Th. J. de Boer, *Tetrahedron* **24**, 1623 (1968).

photochemical nitrosation of hydrocarbons with tertiary alkyl nitrites (*e.g.* *t*-butyl nitrite).[4]

Nevertheless, we observed a nitroxide with a small nitrogen hyperfine splitting (hfs) of 6.2–6.6 gauss when the photochemical nitrosation of a hydrocarbon with *t*-butyl nitrite was carried out in the presence of carbon tetrachloride. Since a *tertiary* nitrite ester was used, it seemed likely that this nitroxide was produced by the scavenging of a radical derived from CCl_4. Because polyhalomethanes like CCl_4 are known to produce polyhalomethyl radicals in free-radical reactions,[5] we proposed that the formation of trichloromethyl alkyl nitroxides II was responsible for the 6.2–6.6 gauss signals.[6]

$$R-N=O \quad + \quad \overset{\bullet}{C}Cl_3 \quad \longrightarrow \quad R-\underset{\underset{O}{|\bullet}}{N}-CCl_3$$

$$\text{I} \qquad\qquad\qquad\qquad\qquad \text{II}$$

The same signals were produced when trichloromethyl radicals were generated from several other sources in the presence of nitrosoalkanes, *i.e.* photolysis of carbon tetrachloride or chloroform in the presence of benzoyl peroxide or di-*t*-butyl peroxyoxalate[7] and photolysis of bromotrichloromethane in benzene.[6]

In the course of our studies the assignment of a trichloromethyl alkyl nitroxide II to the observed ESR spectra has been questioned[8] and in subsequent studies the real spectrum of II with $a_N = 12.73$ gauss, showing a resolved hfs of three chlorine atoms ($a_{Cl} = 2.41$ gauss), has been described.[8–11]

Because the ESR spectra that we have reported[6] showed $a_N = 6.2$–6.6 gauss and $g = 2.0070$, it was suggested that an acyl alkyl nitroxide of unknown origin might be involved.[8] This has stimulated research in this particular field and very recently we published the results mentioned in the following section,[1] almost simultaneously with Holman and Perkins[12] who also reached the conclusion that a new type of acyl alkyl nitroxides was indeed formed.

[5] C. Walling and E. S. Huyser in "Organic Reactions," Vol. 13, Wiley, New York (1963), p. 91.

[6] J. W. Hartgerink, J. B. F. N. Engberts, Th. A. J. W. Wajer and Th. J. de Boer, *Rec. Trav. Chim.* **88**, 481 (1969).

[7] P. D. Bartlett, E. P. Benzing and R. E. Pincock, *J. Amer. Chem. Soc.* **82**, 1762 (1960).

[8] K. Torssell, *Tetrahedron* **26**, 2759 (1970).

[9] I. H. Leaver, G. C. Ramsay and E. Suzuki, *Austr. J. Chem.* **22**, 1891 (1969).

[10] M. J. Perkins, P. Ward and A. Horsfield, *J. Chem. Soc.* (*B*) 395 (1970).

[11] E. G. Janzen, B. R. Knauer, L. T. Williams and W. B. Harrison, *J. Phys. Chem.* **74**, 3025 (1970).

[12] R. J. Holman and M. J. Perkins, *Chem. Commun.* 244 (1971).

RESULTS AND DISCUSSION

In connection with our experiments on the trapping of photochemically generated trichloromethyl radicals by monomeric nitrosoalkanes (*e.g.* 2-nitroso-2-methylpropane), we extended our study to other tri-, di- and monohalomethyl radicals. Suitable sources for these radicals are the polyhalomethanes given in the first column of table 1. Free radical reactions of polyhalomethanes have been studied extensively; especially their peroxide-induced addition to olefins is well understood. In the course of these studies,[5] generalizations could be made regarding the carbon-halogen bond in the polyhalomethane that initially breaks (*cf.* table 1, second column). In all cases nitroxides are observed with a nitrogen hfs of 6.5–7.0 gauss and a *g*-value of 2.0070–2.0072. These values are close to those expected for acyl alkyl nitroxides (a_N = 7.4–8.3 gauss and g = 2.0067–2.0069).[4]

TABLE 1

HFS constants (in gauss) of acyl alkyl nitroxides formed at 20° by photolysis of polyhalomethanes[a] in the presence of 2-nitroso-2-methylpropane

Radical source	Radical precursor	Nitroxide found		a_N	Other hfs	g-value
CCl_4[b]	$^{\bullet}CCl_3$	t-Bu−N−C−Cl (with N−O$^{\bullet}$ and C=O)	(VIIa)[c,d]	6.5	a_{Cl} not resolved	2.0070
CBr_4[b]	$^{\bullet}CBr_3$	t-Bu−N−C−Br (with N−O$^{\bullet}$ and C=O)	(VIIb)[c]	6.7	a_{Br} = 2.0	2.0071
$CHBr_3$[e]	$^{\bullet}CHBr_2$	t-Bu−N−C−H (with N−O$^{\bullet}$ and C=O)	(VIIc)	7.0	a_H = 1.4	2.0071
$CDBr_3$[f]	$^{\bullet}CDBr_2$	t-Bu−N−C−D (with N−O$^{\bullet}$ and C=O)	(VIId)	7.0	a_D not resolved	2.0071
CHI_3[e]	$^{\bullet}CHI_2$	t-Bu−N−C−H (with N−O$^{\bullet}$ and C=O)	(VIIc)	6.9	a_H = 1.4	2.0072
CH_2I_2[e]	$^{\bullet}CH_2I$	t-Bu−N−C−H (with N−O$^{\bullet}$ and C=O)	(VIIc)	6.9	a_H = 1.4	2.0072

[a] Liquid polyhalomethanes were used without cosolvent; (solid) CBr_4 and CHI_3 were dissolved in benzene.

[b] UV irradiation in the presence of benzoyl peroxide and *t*-BuNO (± 5 mg/ml).

[c] Ref. 12 reports a_N = 6.65 gauss for VIIa and a_N = 6.50 gauss and a_{Br} = 2.12 gauss for VIIb.

[d] We have published spectral data of several other adducts of the chlorocarbonyl moiety to various nitrosoalkanes; *cf.* ref. 6.

[e] UV or 680 nm irradiation in the presence of *t*-BuNO (± 5 mg/ml); *cf.* ref. 13.

[f] UV irradiation in the presence of *t*-BuNO (± 5 mg/ml).

[13] Th. A. J. W. Wajer, Thesis, University of Amsterdam (1969), p. 95–97.

We propose initial formation of (poly)halomethyl alkyl nitroxides. Since elimination of a β-halogen atom is generally very easy in free radicals[14] our nitroxides will produce halonitrones III.[15]

The trifluoromethyl alkyl nitroxide (II, X = F) forms an exception and is very stable since C–F bond homolysis is difficult.

Possibly, the reaction from II to III is reversible, since the scavenging of chlorine atoms (generated by *thermal* decomposition of *t*-butyl hypochlorite) by α-phenyl-*N*-*t*-butylnitrone has been reported.[11,16] However, the ESR signal of the derived α-chlorobenzyl *t*-butyl nitroxide was not observed when the chlorine atoms were created by *photolysis* of chlorine in benzene solution;[17] in this case only an acyl nitroxide was present (*i.e.* benzoyl *t*-butyl nitroxide), apparently because the nitrone was consumed in a light-induced reaction.

There may be several possibilities for the formation of acyl alkyl nitroxides from III. Light-induced isomerization of the nitrone III may give the oxaziridine IV.[18] Oxaziridines are active-oxygen compounds as demonstrated by their ability to oxidize *tertiary* amines smoothly to the corresponding *N*-oxides.[19] On the other hand they can also be oxidized (*e.g.* by peracetic acid) to ketones and nitrosoalkanes, presumably *via* N-oxides.[20] Therefore we assume that oxaziridine IV may undergo a bimolecular oxidation-reduction reaction in solution,[21] parallel to other decomposition reactions.[18,19]

[14] E. S. Huyser and R. H. C. Feng, *J. Org. Chem.* **36**, 731 (1971).
[15] The preparation of an α,α-dihalonitrone (*i.e.* α,α-dichloro-*N*-trichloromethyl-nitrone) has recently been reported. This compound was found to be thermally stable (b.p. 164°); *cf.* V. Astley and H. Sutcliffe, *Tetrahedron Letters* 2707 (1971). The fluorine analogue of this nitrone has also been described; *cf.* V. A. Ginsburg, K. N. Smirnov and M. N. Vasil'eva, *J. Gen. Chem. USSR* **39**, 1304 (1969).
[16] E. G. Janzen, *Accounts Chem. Res.* **4**, 31 (1971).
[17] E. G. Janzen and B. J. Blackburn, *J. Amer. Chem. Soc.* **91**, 4481 (1969).
[18] J. S. Splitter and M. Calvin, *J. Org. Chem.* **30**, 3427 (1965).
[19] W. D. Emmons in "Heterocyclic Compounds with three- and fourmembered Rings," Part I, ed. A. Weissberger, Interscience, New York (1964), p. 639.
[20] W. D. Emmons, *ibid.*, p. 645.
[21] In view of the sensitivity of the ESR Spin Trapping technique no conclusion regarding the extent of this oxidation-reduction reaction can be drawn.

Decomposition of V (X = chlorine) into phosgene and RNO cannot explain the formation of VIIa, since we were unable to induce carbon-halogen bond fission in phosgene to yield the chlorocarbonyl radical. For instance, photolysis of a toluene solution of phosgene and RNO does not produce VIIa, neither in the presence of initiators such as benzoyl peroxide. It seems more likely that V undergoes ring-opening and halogen-elimination (possibly concerted) to yield VII. Bluhm and Weinstein[22] report the formation of benzoyl *t*-butyl nitroxide by photolysis of a benzene solution of the relatively stable 2-*t*-butyl-3-phenyloxaziridine. These results suggest that the formation of VII from IV is also light-induced, though no definite conclusions can be drawn, since the halogen atoms in the species of our study may alter the character of the intermediates.

Bromine hfs is observed in the spectrum of VIIb; a total of twelve lines of equal intensity is expected, since bromine has $I = \frac{3}{2}$. A ten line spectrum is actually observed (figure 2), but careful analysis of the spectrum using a small modulation amplitude (figure 3) reveals that the fourth and fifth and also the eighth and ninth line of the expected spectrum overlap in such a way that the ten line signal results.

No chlorine hfs is resolved in the spectrum of VIIa; *cf.* figure 1. This is explained by the smaller nuclear magnetic moment and possibly also by the fact that chlorine is less bulky than bromine, which renders through-space interaction with the π-electron system less important.

Though the nine equivalent γ-hydrogen atoms in the *t*-butyl moiety do not show resolved hfs (the a_H is in the order of 0.1 gauss[23]) they certainly will cause line-broadening and this obscures other small splittings (*e.g.* from chlorine in VIIa). Recently Holman and Perkins[12] have used per-deutero-2-nitroso-2-methylpropane as a scavenger for trichloromethyl radicals. Since the nuclear magnetic moment of deuterium is about 3.3 times smaller than that of hydrogen, the (unresolved) splittings of the nine equivalent γ-deuterium atoms in perdeutero-VIIa will be much smaller

[22] A. L. Bluhm and J. Weinstein, *J. Amer. Chem. Soc.* **92**, 1444 (1970).
[23] Th. A. J. W. Wajer, Thesis, University of Amsterdam (1969), p. 131.

than 0.1 gauss and the linewidth of the spectrum is therefore reduced to such an extent that the chlorine hfs becomes visible (a_{Cl} = 0.45 gauss).

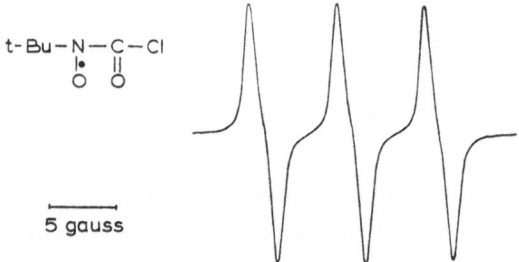

Figure 1. ESR spectrum of nitroxide VIIa in CCl_4 at 20°.

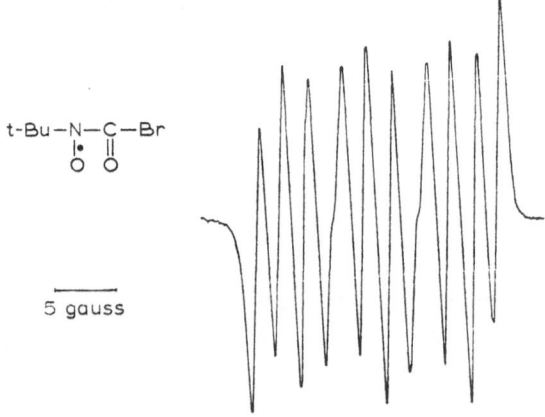

Figure 2. ESR spectrum of nitroxide VIIb in benzene at 20°.

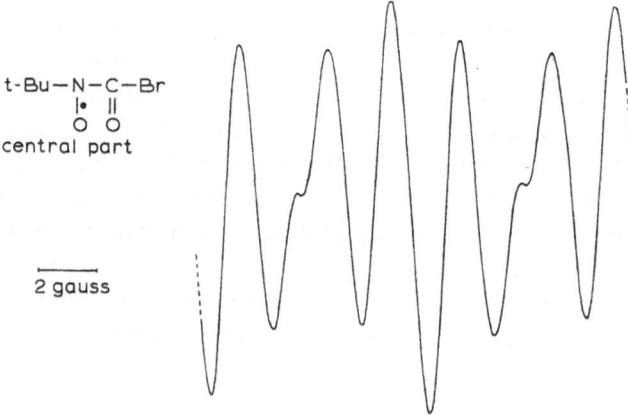

Figure 3. Central part of the ESR spectrum of nitroxide VIIb.

24

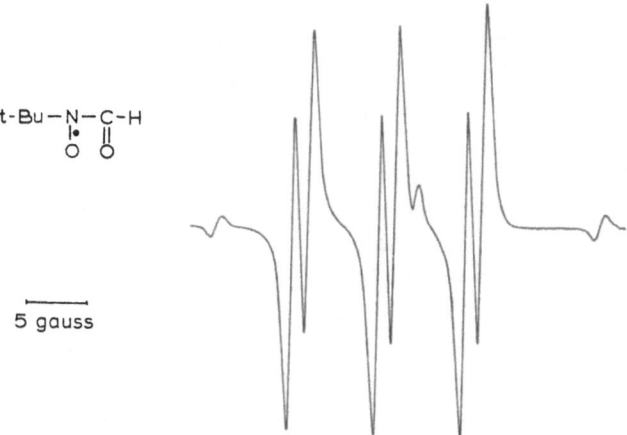

Figure 4. ESR spectrum of nitroxide VIIc in CHBr$_3$ at 20°.
The impurity is di-t-butyl nitroxide.

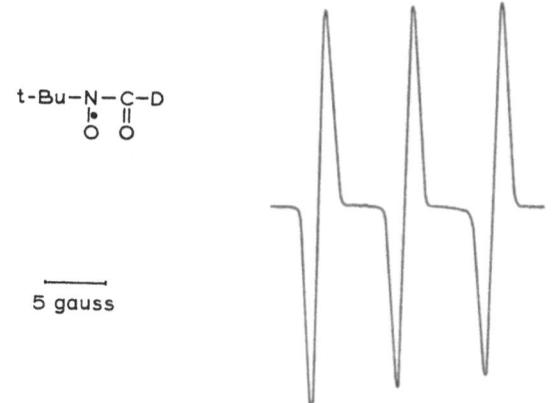

Figure 5. ESR spectrum of nitroxide VIId in CDBr$_3$ at 20°.

The acyl alkyl nitroxides VII are stable for days at 20°, as judged from the slow disappearance of the ESR signals. The stability of VIIa and VIIb has also been noted by Holman and Perkins,[12] who further report that these species undergo rapid reactions with nucleophiles, $e.g.$ both VIIa and VIIb immediately produce the *same* (new) nitroxide when ethanol is added to the sample. Apparently the halogen atom has been replaced by the ethoxy moiety in both cases.

When dihalomethyl radicals are trapped by t-BuNO, presumably similar reactions take place leading to the formation of VIIc (figure 4). The structure of VIIc is supported by a preparation *via* an alternative route, *i.e.* hydrogen

25

abstraction from formaldehyde by *t*-butoxy radicals (generated thermally from di-*t*-butyl peroxyoxalate[7]) and addition of the formyl radical to I. When the deuterodibromomethyl radical is generated the hydrogen splitting, of course, disappears while the much smaller deuterium hfs remains unresolved in VIId (figure 5).

Nitroxide VIIc is also formed even when *mono*halomethyl radicals are generated in the presence of I. The initially formed monohalomethyl *t*-butyl nitroxide apparently loses halogen to give the halogen-free nitroxide VI (X = H). Abstraction of hydrogen (*e.g.* by oxaziridine[22]) leads to VIIc. However, fragmentation of VI (X = H) to I and formaldehyde cannot be excluded in this case, since we mentioned already that VIIc can be generated easily from these species.

Bluhm and Weinstein[22] have shown that various stable nitrones (*e.g.* α-phenyl-*N*-*t*-butylnitrone) yield acyl alkyl nitroxides upon irradiation. These results can be explained by either of the above mechanisms.

Besides the routes indicated the possibility cannot be excluded that certain nitroxides form addition products with nitrones, which decompose into acyl alkyl nitroxides as reported by Holman and Perkins.[12]

The decomposition of the intermediate is somewhat similar to the one proposed in the bimolecular reactions of *t*-butyl *p*-halophenyl nitroxides.[24]

EXPERIMENTAL

Materials

All polyhalomethanes were pure commercial samples; the deuterobromoform ("*Merck*") had at least 99% deuterium content. The synthesis of 2-nitroso-2-methylpropane has been described in chapter 2.

ESR Spin Trapping (for technical details see chapter 2)

A small amount of 2-nitroso-2-methylpropane was dissolved in the polyhalomethane (roughly 5 mg per ml); in some cases (*cf.* table 1) a suitable initiator (benzoyl peroxide) was also added. Benzene was used as the solvent for the

---

[24] A. R. Forrester, private communication.

solid polyhalomethanes (CBr_4 and CHI_3). Nitrogen was passed through the samples to remove oxygen.

Irradiation for a short time produced the acyl alkyl nitroxides. Usually no other signals were present in the spectrum; occasionally a small amount of di-*t*-butyl nitroxide was formed, which was easily identified in the spectra (see for instance figure 4). Some of the spectra in table 1 have already been described by Wajer[13] and can now be reinterpreted in the light of our latest findings, *i.e.* that low splitting constants ($\pm$ 6.7 gauss) indicate the presence of acyl alkyl nitroxides.

PHOTOCHEMICALLY INITIATED REACTIONS OF SUBSTITUTED 1,3-DIOXOLANES AND 1,3-OXATHIOLANES IN CFCl₃

ESR STUDY AND MECHANISM OF RING-FISSION

Abstract – The photochemical reactions of some 2-alkyl-1,3-dioxolanes and 2-alkyl-1,3-oxathiolanes in CFCl₃ in the presence of benzophenone yield exclusively the *open* 2-chloroethyl carboxylic esters and *S*-2-chloroethyl thiocarboxylic esters respectively. Photochemically excited benzophenone abstracts the hydrogen atom at carbon between the heteroatoms from the substrate to give intermediate *cyclic* (thio)acetal radicals which can be trapped efficiently by 2-nitroso-2-methylpropane in inert solvents. The resulting nitroxides are identified by their ESR hfs constants. No ring-opened (thio)ester radicals could be trapped. The course of photolysis of optically active 2RS,4R-(−)-2-methyl-4-phenyl-1,3-dioxolane and other (racemic) 2,4-disubstituted-1,3-dioxolanes supports a mechanism in which a cyclic radical abstracts halogen from the solvent to form an intermediate cyclic chloro-(thio)acetal. Heterolytic cleavage of the new C–Cl bond gives the well stabilized cyclic carbonium ion and chloride anion. Nucleophilic attack of chloride ion at the C-4 or C-5 carbon atom (involving inversion for a chiral C-4) leads to ring rupture and formation of the final product.

INTRODUCTION

Free radical reactions of acetals, induced by thermally generated alkoxy radicals[2-6] or by photoactivated ketones[7-9] have been studied by several groups of investigators. A variety of reactions has been detected, often involving hydrogen abstraction from the carbon adjacent to both oxygen atoms (as illustrated for 2-alkyl-1,3-dioxolane I) producing a cyclic acetal radical II which usually isomerizes to the ester radical III. This ring-opened

---

[1] The contents of this chapter are nearly identical with a full paper: J. W. Hartgerink, L. C. J. van der Laan, J. B. F. N. Engberts and Th. J. de Boer, *Tetrahedron* 27, 4323 (1971).

[2] L. P. Kuhn and C. Wellman, *J. Org. Chem.* 22, 774 (1957).

[3] E. S. Huyser, *J. Org. Chem.* 25, 1820 (1960).

[4] E. S. Huyser and Z. Garcia, *J. Org. Chem.* 27, 2716 (1962).

[5] B. Maillard, M. Cazaux and R. Lalande, *Bull. Soc. Chim. France* 467 (1971).

[6] P. Marche and D. Lefort, *C. R. Acad. Sci. Paris, Ser. C* 269, 717 (1969).

[7] D. Elad and R. D. Youssefyeh, *Tetrahedron Letters* 2189 (1963).

[8] I. Rosenthal and D. Elad, *J. Org. Chem.* 33, 805 (1968).

[9] H. E. Seyfarth, A. Hesse and H. Pastohr, *Z. Chem.* 9, 150 (1969).

radical may abstract a hydrogen atom from the solvent to yield the carboxylic ester IV.[10]

$$
\underset{\text{I}}{R-\overset{\overset{\displaystyle H}{|}}{\underset{\underset{\displaystyle O-CH_2}{|}}{C}}\!\!\begin{array}{c}O-CH_2\\ |\\ \end{array}}
\longrightarrow
\underset{\text{II}}{R-\overset{\overset{\displaystyle \cdot}{|}}{\underset{\underset{\displaystyle O-CH_2}{|}}{C}}\!\!\begin{array}{c}O-CH_2\\ |\\ \end{array}}
\longrightarrow
\underset{\text{III}}{R-\overset{\displaystyle O}{\underset{\underset{\displaystyle O-\dot{C}H_2}{|}}{C}}\!\!\begin{array}{c}\dot{C}H_2\\ |\\ \end{array}}
\longrightarrow
\underset{\text{IV}}{R-\overset{\displaystyle O}{\underset{\underset{\displaystyle O-CH_2-CH_3}{}}{C}}}
$$

Hydrogen abstraction at C-2 from 2,4-disubstituted-1,3-dioxolanes may
result in β-scission to give predominantly the n-alkyl ester derived from the
more stable secondary radical.[4,9] In several cases cyclic acetal radicals have
been trapped by alkenes to give the 1:1 addition product, most efficiently
by diethyl maleate under photolytic conditions; no trace of the ring-opened
ester radical can be trapped in this case.[8] At higher temperatures both the
cyclic radical and the ester radical are trapped by 1-octene. It seems that
the (homolytic) ring-opening reaction is temperature dependent, since the
amount of trapped ester radicals is much smaller at 90° than at 160°.[5]

The purpose of the present investigation was twofold. First of all to study
the effect of substituents in 1,3-dioxolanes and 1,3-oxathiolanes on the
propensity for formation of open *versus* closed radicals by means of the
ESR Spin Trapping technique[11,12] with 2-nitroso-2-methylpropane V as a
scavenger. Secondly, to elucidate the reaction mechanism (homolysis and/or
heterolysis) of the benzophenone-induced photochemical reactions of substituted 1,3-dioxolanes and 1,3-oxathiolanes in $CFCl_3$ as a solvent.

RESULTS AND DISCUSSION

An ESR study of radicals derived from substituted 1,3-*dioxolanes*
We have used 2-nitroso-2-methylpropane V as a highly efficient trap in
the free radical reactions of a number of substituted 1,3-dioxolanes under
various conditions. In benzophenone-induced photochemical reactions and
in thermal reactions with *t*-butoxy radicals[13] as an initiator the same
nitroxides VI, derived from the *cyclic* acetal radicals are observed; these

---

[10] These free radical reactions must be carried out under nitrogen, because in the
presence of oxygen mainly hydroperoxides are obtained; *cf.* ref. 9.

[11] J. W. Hartgerink, J. B. F. N. Engberts, Th. A. J. W. Wajer and Th. J. de Boer,
Rec. Trav. Chim. **88**, 481 (1969).

[12] C. Lagercrantz and S. Forshult, *Acta Chem. Scand.* **23**, 811 (1969).

[13] Formed by thermal decomposition of di-*t*-butyl peroxyoxalate (TBPO) at room
temperature; *cf.* P. D. Bartlett, E. P. Benzing and R. E. Pincock, *J. Amer. Chem. Soc.* **82**,
1762 (1960).

nitroxides are sufficiently stable to be studied at room temperature by ESR in benzene solution.

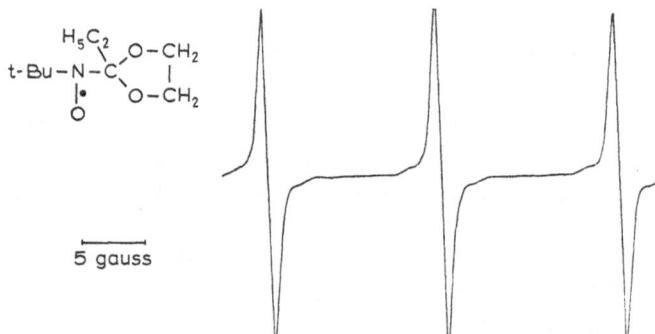

Especially under photochemical conditions the trapping agent itself gives rise to the formation of di-*t*-butyl nitroxide; the signal of this radical is sometimes so strong that the signal from VI is hardly visible. The structures of the nitroxides VI are readily deduced from the observed hfs constants. In most cases only a nitrogen splitting is observed with a_N values of 13–14 gauss (table 1) which is somewhat lower than usually found in di-alkyl nitroxides ($a_N = 15$–16 gauss). The slightly reduced spin density at nitrogen will be the result of the inductive effect of the oxygen atoms attached to the α-carbon atom. This is compatible with the results obtained by Torssell,[14] indicating that the inductive effect of β-carbonyl groups in nitroxides is of the same magnitude as that of β-alkoxy groups. The lowering of the a_N value by approximately 1.7 gauss in his nitroxides with two β-carbonyl groups is indeed comparable with the lowering that we observe in nitroxides VI.

In benzene no nitroxides derived from the open ester radical are observed, not even when four methyl groups are attached at C-4 and C-5 to facilitate the ring-opening. The ESR spectrum of the nitroxide derived from 4,4,5,5-tetramethyl-1,3-dioxolane shows an additional splitting due to one β-hydrogen atom, as expected for the trapped cyclic radical (*cf.* figure 2).

Figure 1. ESR spectrum of the nitroxide which is formed in benzene at 20° by hydrogen abstraction from 2-ethyl-1,3-dioxolane in the presence of 2-nitroso-2-methylpropane.

[14] K. Torssell, *Tetrahedron* **26**, 2759 (1970).

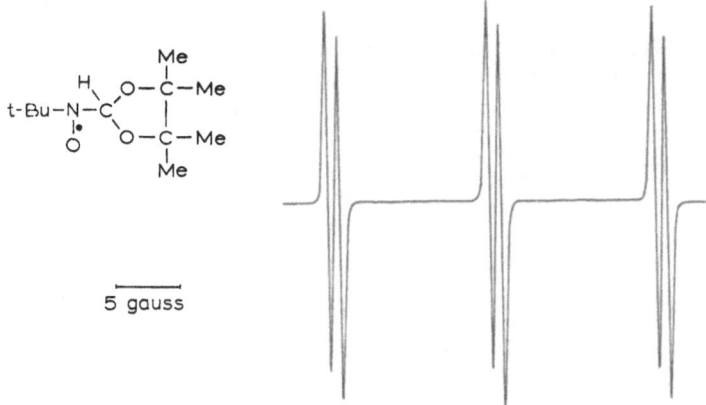

Figure 2. ESR spectrum of the nitroxide which is formed in benzene at 20° by hydrogen abstraction from 4,4,5,5-tetramethyl-1,3-dioxolane in the presence of 2-nitroso-2-methylpropane.

In the case of unsubstituted 1,3-dioxolane a complex ESR spectrum is obtained, apparently due to nitroxides derived from a mixture of acetal radicals formed by hydrogen abstraction at C-2 and C-4.[15] Apparently, selective hydrogen abstraction at C-2 only takes place when a substituent at this position provides extra stabilization of the radical.

In table 1 ESR data of two nitroxides derived from acyclic acetals have been included for comparison.

The stability and geometry of oxygen-conjugated radicals is under active discussion in the recent literature.[16,17] Norman[18,19] has obtained strong evidence that these radicals possess a non-planar structure by measuring ^{13}C hfs constants. It is interesting to note that for 2-cyclopropyl-1,3-dioxolane only the ESR spectrum from a nitroxide with an intact cyclopropyl ring is observed. This is somewhat surprising because α-cyclopropyl carbinyl radicals usually undergo a rapid ring opening[20] unless there is strong resonance stabilization of the radical center.[21,22] It may tentatively be proposed

[15] The non-specific hydrogen abstraction from this compound has been noted before; cf. ref. 5.

[16] A. Ohno and Y. Ohnishi, *Tetrahedron Letters* 4405 (1969).

[17] A. Hudson and K. D. J. Root, *Tetrahedron* 25, 5311 (1969).

[18] A. J. Dobbs, B. C. Gilbert and R. O. C. Norman, *Chem. Commun.* 1353 (1969); idem, *J. Chem. Soc. (A)* 124 (1971).

[19] R. O. C. Norman, *Chem. in Brit.* 6, 66 (1970).

[20] The cyclopropyl carbinyl radical is stable at −150° and gives β-scission at −100° to form the allyl carbinyl radical; cf. J. K. Kochi, P. J. Krusic and D. R. Eaton, *J. Amer. Chem. Soc.* 91, 1877 (1969).

[21] D. C. Neckers, A. P. Schaap and J. Hardy, *J. Amer. Chem. Soc.* 88, 1265 (1966).

[22] J. C. Martin, J. E. Schultz and J. W. Timberlake, *Tetrahedron Letters* 4629 (1967).

31

TABLE 1

HFS constants (in gauss) of nitroxides formed in benzene at 20°
by hydrogen abstraction from cyclic and acyclic acetals in the
presence of 2-nitroso-2-methylpropane

Trapped radical R· from parent acetal RH	t-Bu(R)NO·	
	a_N	$a_{H\beta}$
CH$_3$–Ċ⟨O⟩ (dioxolane)	14.1	–
C$_2$H$_5$–Ċ⟨O⟩ (dioxolane)	13.9	–
n–C$_3$H$_7$–Ċ⟨O⟩ (dioxolane)	14.1	–
▷–Ċ⟨O⟩ (dioxolane)	14.1	–
Ø–CH$_2$–Ċ⟨O⟩ (dioxolane)	13.8	–
Ø–Ċ⟨O⟩ (dioxolane)	14.1	–
(thiophene)–Ċ⟨O⟩ (dioxolane)	14.0	–
H–Ċ⟨O⟩ (dimethyldioxolane)	13.2	0.94 (1H)
H–Ċ(OCH$_3$)$_2$	13.5	1.90 (1H)
CH$_3$–Ċ(OCH$_3$)$_2$	13.6	–

that the decreased propensity for cyclopropyl ring opening for the radical derived from 2-cyclopropyl-1,3-dioxolane might be the result of a non-planar geometry.

In the following section the photolyses of some substituted 1,3-dioxolanes in the chlorine-donating solvent CFCl$_3$ are described. Although ring-opened products are finally obtained, we will show that in agreement with the ESR experiments, no ring-opened radicals are involved in the photochemically initiated reactions of 1,3-dioxolanes.

Reactions of substituted 1,3-dioxolanes with excited benzophenone in $CFCl_3$

The photochemically induced reactions of four 2-alkyl-1,3-dioxolanes were studied in $CFCl_3$ as the solvent[23] with benzophenone as the initiator[24] (table 2).

In all cases a smooth conversion into a 2-chloroethyl carboxylic ester VII is observed. This can be explained by either of the following two mechanisms A and B.

SCHEME 1. MECHANISM A (HYPOTHETICAL)

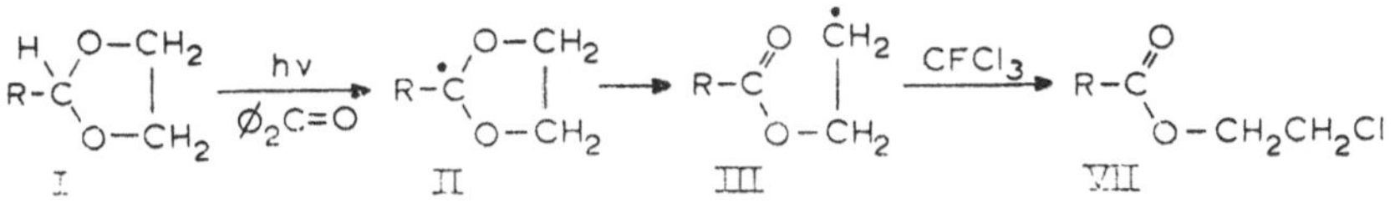

$$
\underset{\text{I}}{R-\overset{\displaystyle H}{\underset{\displaystyle}{C}}\Big\langle\begin{smallmatrix}O-CH_2\\ \;\\ O-CH_2\end{smallmatrix}} \xrightarrow[\phi_2C=O]{h\nu} \underset{\text{II}}{R-\overset{\displaystyle \bullet}{C}\Big\langle\begin{smallmatrix}O-CH_2\\ \;\\ O-CH_2\end{smallmatrix}} \longrightarrow \underset{\text{III}}{R-\overset{\displaystyle O}{\underset{\displaystyle}{C}}\Big\langle\begin{smallmatrix}\overset{\bullet}{C}H_2\\ \;\\ O-CH_2\end{smallmatrix}} \xrightarrow{CFCl_3} \underset{\text{VII}}{R-\overset{\displaystyle O}{\underset{\displaystyle}{C}}\diagdown_{O-CH_2CH_2Cl}}
$$

The ESR study did not provide any evidence for β-scission of cyclic acetal radicals II and this renders mechanism A (scheme 1) *a priori* improbable.

In mechanism B the formation of a 2-chloro-2-alkyl-1,3-dioxolane VIII is assumed, apparently *via* abstraction of a chlorine atom from the solvent[25] by the *cyclic* acetal radical II.

SCHEME 2. MECHANISM B

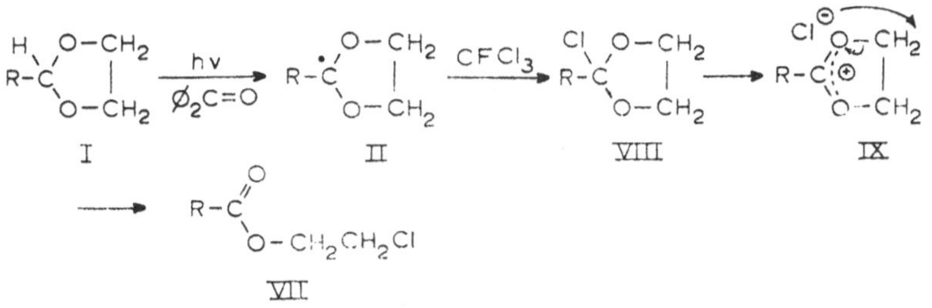

$$
\underset{\text{I}}{R-\overset{H}{\underset{}{C}}\Big\langle\begin{smallmatrix}O-CH_2\\ O-CH_2\end{smallmatrix}} \xrightarrow[\phi_2C=O]{h\nu} \underset{\text{II}}{R-\overset{\bullet}{C}\Big\langle\begin{smallmatrix}O-CH_2\\ O-CH_2\end{smallmatrix}} \xrightarrow{CFCl_3} \underset{\text{VIII}}{R-\overset{Cl}{\underset{}{C}}\Big\langle\begin{smallmatrix}O-CH_2\\ O-CH_2\end{smallmatrix}} \longrightarrow \underset{\text{IX}}{R-\overset{Cl^{\ominus}}{\underset{}{C}}^{\oplus}\Big\langle\begin{smallmatrix}\overset{\oplus}{O}-CH_2\\ O-CH_2\end{smallmatrix}}
$$

$$
\longrightarrow \underset{\text{VII}}{R-\overset{O}{\underset{}{C}}\diagdown_{O-CH_2CH_2Cl}}
$$

Such cyclic chloro-acetals are known to be very reactive species that have never been isolated. Although Baker *et al.*[26] reported the preparation of 2-chloro-1,3-dioxolane by photochemical chlorination of 1,3-dioxolane,

---

[23] All studies reported in the literature have been performed in hydrogen donating media, either the pure 1,3-dioxolane itself or with an alcohol as the solvent.

[24] No reaction occurred in the absence of benzophenone; benzpinacol was found as a byproduct.

[25] The fate of the $CFCl_2$ radical thus produced remains uncertain. It may participate in the hydrogen abstraction reaction to form $CHFCl_2$ (b.p. 9°) or it dimerizes to $Cl_2FCCFCl_2$ (b.p. 93°). A careful search has been made for these compounds (GLC analysis) and neither compound could be detected.

[26] W. Baker and A. Shanon, *J. Chem. Soc.* 1598 (1933).

33

TABLE 2

Photolyses of substituted 1,3-dioxolanes and 1,3-oxathiolanes in $CFCl_3$ in the presence of benzophenone

Starting material	Reaction product(s)	Yield (%)	b.p.	NMR spectral data[a]
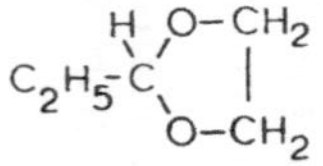 2-ethyl-1,3-dioxolane (H, C_2H_5 at C-2)	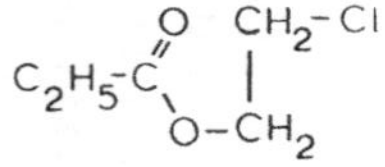 C_2H_5–C(=O)–O–CH_2–CH_2–Cl	35	67–69°/30 mm	1.16 (t, 3H), 2.38 (q, 2H), 3.70 (t, 2H), 4.34 (t, 2H).
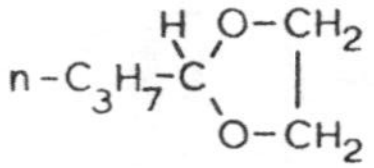 2-n-propyl-1,3-dioxolane (H, n-C_3H_7 at C-2)	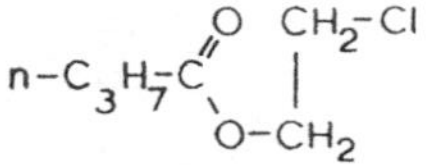 n-C_3H_7–C(=O)–O–CH_2–CH_2–Cl	62	74–76°/30 mm	0.95 (t, 3H), 1.65 (m, 2H), 2.38 (t, 2H), 3.64 (t, 2H), 4.26 (t, 2H).
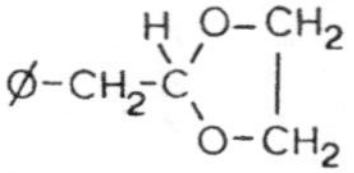 2-benzyl-1,3-dioxolane (ϕ–CH_2 at C-2)	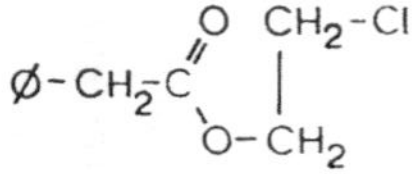 ϕ–CH_2–C(=O)–O–CH_2–CH_2–Cl	54[b]	160°/15 mm	3.58 (t, 2H), 3.63 (s, 2H), 4.29 (t, 2H), 7.29 (s, 5H).
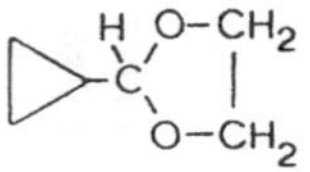 1,4-dioxaspiro[4.2]heptane (cyclopropane-spiro dioxolane)	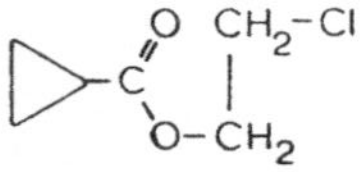 cyclopropane–C(=O)–O–CH_2–CH_2–Cl	50	41–42°/4 mm	0.91 (m, 4H), 1.60 (m, 1H), 3.63 (t, 2H), 4.27 (t, 2H).
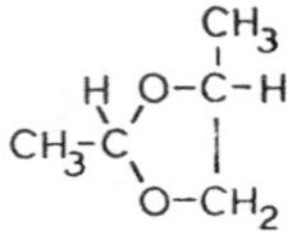 2-methyl-4-methyl-1,3-dioxolane (CH_3, H at C-4; CH_3, H at C-2)	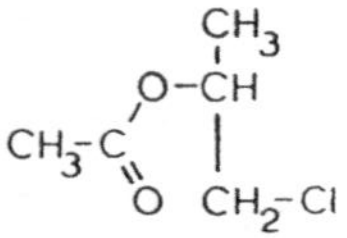 CH_3–C(=O)–O–CH(CH_3)–CH_2–Cl	49	47–51°/15 mm	1.33 (d, 3H), 2.05 (s, 3H), 3.57 (d, 2H), 5.13 (m, 1H).
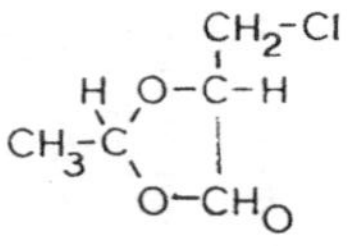 2-methyl-4-chloromethyl-1,3-dioxolane (CH_2Cl, H at C-4; CH_3, H at C-2)	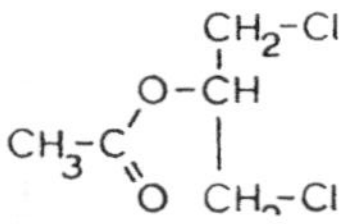 CH_3–C(=O)–O–CH(CH_2Cl)–CH_2O	44	97–98°/22 mm	2.13 (s, 3H), 3.75 (d, 4H), 5.18 (qnt, 1H).

34

Starting material	Product	Yield / rotation	b.p.	NMR (δ, ppm)
2 RS,4R-(−)[c,d] ϕ-C4-H, CH$_3$-C2, O-CH$_2$ (1,3-dioxolane)	S-(+) ϕ HC-Cl, CH$_3$-C(=O)-O-CH$_2$	52.6[a]	113–115°/2 mm	2.02 (s, 3H), 4.45 (d, 2H), 5.09 (t, 1H), 7.38 (s, 5H).
	R-(−) ϕ-C-H, CH$_3$-C(=O)-O-CH$_2$-Cl	4.0[a]	119°/1.5 mm	2.09 (s, 3H), 3.74 (d, 2H), 6.00 (t, 1H), 7.36 (s, 5H).
C$_2$H$_5$, H, O-CH$_2$, S-CH$_2$ (oxathiolane)	C_2H_5-C(=O)-S-CH$_2$, CH$_2$-Cl	22[b,d]	82°/15 mm	1.19 (t, 3H), 2.58 (q, 2H), 3.17 and 3.58 (A$_2$B$_2$, 4H).
n-C$_3$H$_7$, H, O-CH$_2$, S-CH$_2$	n-C$_3$H$_7$-C(=O)-S-CH$_2$, CH$_2$-Cl	38[b,d]	90°/16 mm	0.95 (t, 3H), 1.69 (m, 2H), 2.58 (t, 2H), 3.23 and 3.53 (A$_2$B$_2$, 4H).
i-C$_3$H$_7$, H, O-CH$_2$, S-CH$_2$	i-C$_3$H$_7$-C(=O)-S-CH$_2$, CH$_2$-Cl	14[b,d]	88°/15 mm	1.18 (d, 6H), 2.75 (spt, 1H), 3.17 and 3.62 (A$_2$B$_2$, 4H).

[a] Chemical shifts in ppm (δ-values). Spectra of S-2-chloroethyl thioesters in CCl$_4$; the other spectra in CDCl$_3$. s = singlet, d = doublet, t = triplet, q = quartet, qnt = quintet, spt = septet, m = multiplet.
[b] Yield based on consumed starting material.
[c] Specific rotations of starting material and products: cf. Experimental.
[d] New compounds.

35

Bagans et al.[27] repeated this experiment more recently and isolated only the isomeric 2-chloroethyl formate. One cyclic chloro-acetal (i.e. 2-chloro-2-dichloromethyl-1,3-dioxolane) has been prepared[28] at $-60°$ as judged by its low-temperature NMR spectrum.[29] Since a carbonium ion is strongly stabilized by two neighbouring oxygen atoms the chloro-acetal VIII will easily give heterolytic cleavage of the C–Cl bond. The oxygen atoms with fractional positive charges will facilitate an S_N2 reaction of chloride ion at the neighbouring C-4 carbon atom to yield 2-chloroethyl carboxylic ester VII (scheme 2).

In order to substantiate the occurrence of mechanism B the photolyses of some appropriately substituted chiral 1,3-dioxolanes were carried out in $CFCl_3$ (table 2). For this purpose we synthesized optically active 2RS,4R-(−)-2-methyl-4-phenyl-1,3-dioxolane X. Starting material was R-(−)-mandelic acid; reduction with $LiAlH_4$ yielded R-(−)-1-phenylethane-1,2-diol XIII. Acid catalyzed reaction of XIII with acetaldehyde afforded 2RS,4R-(−)-2-methyl-4-phenyl-1,3-dioxolane (X). Compound X was optically pure,[30] because acid catalyzed hydrolysis yielded R-(−)-XIII with nearly the same rotation as the starting diol (scheme 3).

SCHEME 3. HYDROLYSIS OF OPTICALLY ACTIVE X[31]

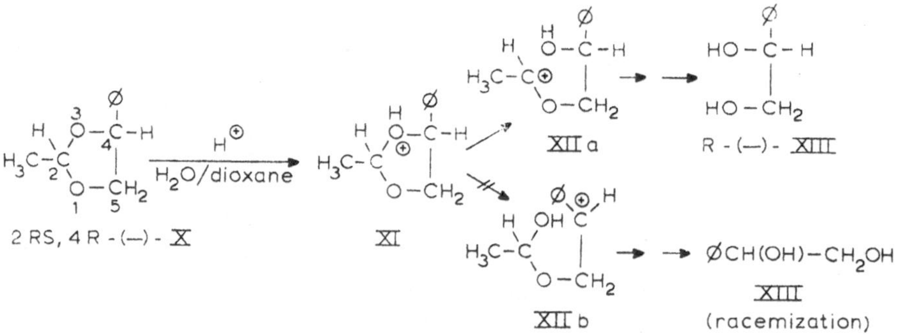

This shows inter alia that a carbonium ion is stabilized much better by a neighbouring oxygen atom than by a phenyl group, because rupture of the O-3–C-4 bond would have caused a fair amount of racemization.

[27] H. Bagans and L. Domaschke, Chem. Ber. 91, 653 (1958).
[28] A few open chloro-acetals have been prepared via other routes; cf. J. W. Scheeren, Tetrahedron Letters 5613 (1968).
[29] H. Gross, J. Freiberg and B. Costisella, Chem. Ber. 101, 1250 (1968).
[30] Optically pure at chiral center C-4. In practice X is a mixture of diastereoisomers due to the other asymmetric center C-2. However, the configuration at C-2 is not relevant in the mechanism of the photolysis of X.
[31] Only protonation of the O-3 oxygen atom is exemplified in this scheme.

Stereochemistry of ester formation

When 2RS,4R-(−)-2-methyl-4-phenyl-1,3-dioxolane X is photolysed in the presence of benzophenone in CFCl₃, S-(+)-2-acetoxy-1-chloro-1-phenyl-ethane XVII and R-(−)-1-acetoxy-2-chloro-1-phenylethane XVIII are formed in a ratio 93:7. Both products are new compounds and were also synthesized *via* independent routes (*cf.* Experimental). The β-chloro-ester XVII was found to be formed with 100% inversion of configuration. This strongly suggests that an S_N2 reaction has occurred at C-4; this is in agreement with the mechanism involving the cyclic chloro-acetal XV (scheme 4).[32] Since the carbonium ion in XVI is so well stabilized by two neighbouring oxygens, it has no tendency for heterolytic scission of the O-3–C-4 bond to form a (planar) benzylic carbonium ion and this explains why no racemic XVII is found. Apparently the only possibility for heterolytic ring opening is a simultaneous attack of the chloride ion at C-4 to produce XVII with inversion of configuration. Chiral center C-4 is not involved in the reaction leading to compound XVIII, which is therefore formed with retention of configuration.

SCHEME 4. PHOTOLYSIS OF OPTICALLY ACTIVE X

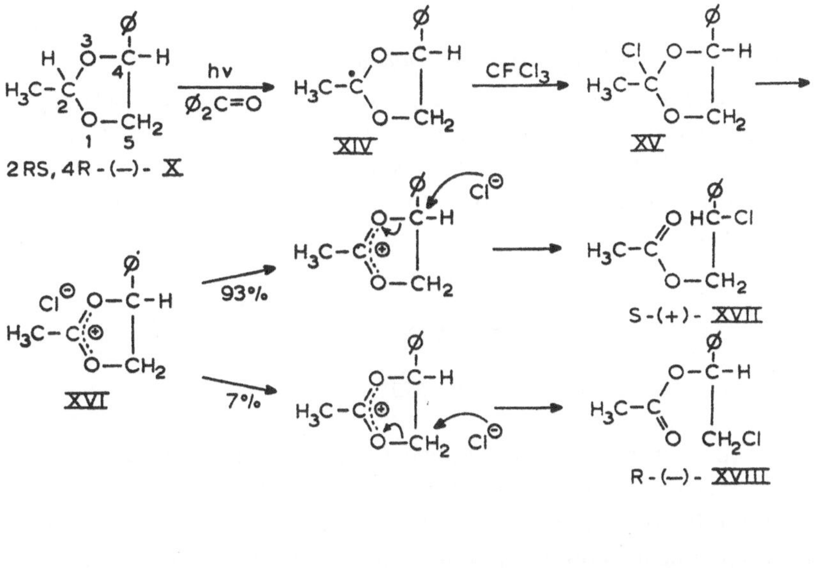

[32] Of course, our experimental data do not provide information about the details of the formation of XVI from XIV. However, the sequence depicted in scheme 4, *i.e.* chlorine abstraction from the solvent to give XV followed by ionization to XVI seems most plausible.

When the phenyl group in X is replaced by methyl or chloromethyl as in XIX, the chloride ion attacks the *least* substituted position in the intermediary carbonium ion to give the ester XXI, while the isomeric ester derived from chloride attack at C-4 is not formed at all.

SCHEME 5. PHOTOLYSIS OF 2,4-DIALKYL-1,3-DIOXOLANES

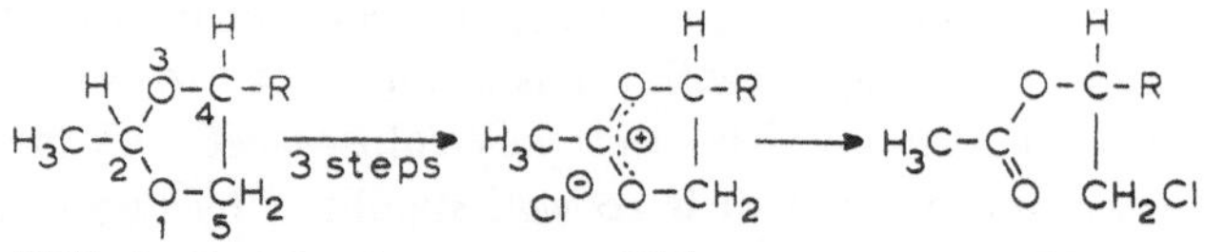

Apparently steric factors are relatively important, whereas in the case with a phenyl substituent electronic factors determine the product selectivity.

The same type of selectivity was recently reported by Gelas et al.[33] for the reaction of 2,4-dialkyl-1,3-dioxolanes with N-bromosuccinimide; again only the least substituted position (C-5) is attacked (by bromide) to form a bromo ester analogous to XXI. The formation of a cyclic bromo-acetal followed by heterolytic ring opening (S_N2 reaction) might well explain these experimental results.

Radical reactions of 2-alkyl-1,3-oxathiolanes

Cyclic thioacetal radicals can be generated by hydrogen abstraction at C-2 from 2-alkyl-1,3-oxathiolanes XXIII by t-butoxy radicals. Trapping with 2-nitroso-2-methylpropane produces nitroxides XXII; the a_N values of these radicals (table 3) are in the same range as those derived from cyclic acetal radicals.

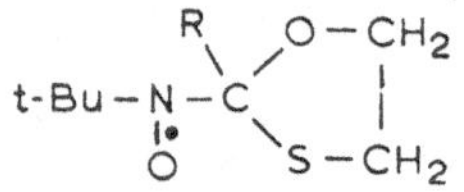

XXII

We have photolysed several 2-alkyl-1,3-oxathiolanes XXIII in $CFCl_3$ in the presence of benzophenone (table 2). The reactions are considerably slower than the photolyses of 2-alkyl-1,3-dioxolanes when carried out under similar experimental conditions and are highly selective. Only the S-2-chloroethyl thioester XXV and no trace of the O-2-chloroethyl thioester is formed.

[33] J. Gelas and S. Michaud, *C. R. Acad. Sci. Paris, Ser. C* **270**, 1614 (1970).

TABLE 3

HFS constants (in gauss) of nitroxides formed in benzene at 20°
by hydrogen abstraction from 2-alkyl-1,3-oxathiolanes in the
presence of 2-nitroso-2-methylpropane

Trapped radical R· from parent acetal RH	t-Bu(R)NO· a_N
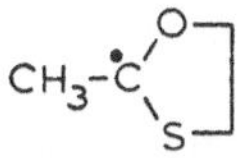	14.4
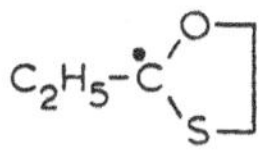	14.3
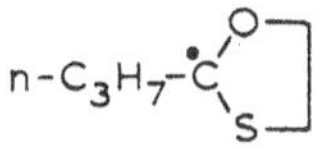	14.3
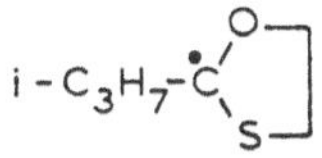	14.3

Assuming a similar mechanism for the photolyses of 2-alkyl-1,3-oxathiolanes as for substituted 1,3-dioxolanes, the observed reaction products can be explained by the fact that a carbonium ion is much better stabilized by a neighbouring oxygen than by a neighbouring sulfur atom.[34] Therefore resonance structure XXIVb is the more important one, which makes the carbon atom attached to oxygen more susceptible for nucleophilic attack than the carbon atom attached to sulfur.

SCHEME 6. PHOTOLYSIS OF 2-ALKYL-1,3-OXATHIOLANES

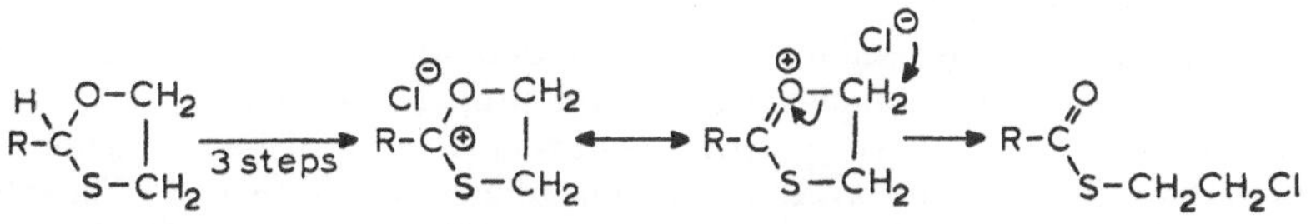

Another factor which may contribute to the observed specific ring opening is the higher stability of S-alkyl thioesters as compared with O-alkyl thioesters.[35]

[34] C. C. Price and S. Oae, "Sulfur Bonding," Ronald Press, New York (1962), p. 10.
[35] T. L. Cottrell, "The Strengths of Chemical Bonds," Butterworths, London (1958).

EXPERIMENTAL

All melting points and boiling points are uncorrected. IR spectra were recorded with a Unicam SP 200 spectrometer. NMR spectra were determined on a Varian A-60 or A-60D instrument using TMS as an internal standard ($\delta = 0$) and CDCl$_3$ or CCl$_4$ as solvent. Optical rotations were measured on a Zeiss LEP polarimeter in a 1 dm tube at 20°, concentrations are in g/100 ml. ESR spectra were taken on a Varian E-3 apparatus fitted with an optical transmission cavity; the light source was a Philips SP 500 W super high pressure mercury lamp. A similar light source was used for the photolyses, which were carried out under nitrogen at 0° in a Pyrex vessel under continuous stirring. Microanalyses were carried out by Mr. H. Pieters in our laboratory.

Materials

All 2-alkyl-1,3-dioxolanes and 2,4-dialkyl-1,3-dioxolanes were prepared from the corresponding aldehydes and diols, using p-TsOH as a catalyst and benzene as the solvent; water was removed azeotropically. All 2-alkyl-1,3-oxathiolanes were prepared in a similar way from the aldehydes and 2-mercapto-ethanol. 4,4,5,5-Tetramethyl-1,3-dioxolane was prepared according to Leutner.[36]

2RS,4R-(−)-2-methyl-4-phenyl-1,3-dioxolane (X)

Reduction of 10.0 g of R-(−)-mandelic acid[37] in 300 ml of dry ether with 6.5 g of LiAlH$_4$ in 150 ml of ether gave 6.1 g of R-(−)-1-phenylethane-1,2-diol, which was crystallized from ether-light petroleum 1:1, m.p. 66.0–68.5°, [α]$_D$ −42.0° (c, 6.5, 96% ethanol).[38]

Water was removed azeotropically from a mixture of 5.8 g of R-(−)-1-phenyl-ethane-1,2-diol, 1.84 g of paraldehyde, 0.20 g of p-TsOH and 150 ml of benzene in the course of 3 hrs. After drying (K$_2$CO$_3$) and removal of the solvent, distillation afforded 5.1 g of 2RS,4R-(−)-2-methyl-4-phenyl-1,3-dioxolane, b.p. 115–116°/21 mm, [α]$_D$ −69.5°, [α]$_{578}$ −73.1° (c, 2.4, cyclohexane). Anal. Found: C, 73.10; H, 7.49; C$_{10}$H$_{12}$O$_2$ requires: C, 73.15; H, 7.37%.

Photolyses of substituted 1,3-dioxolanes and 1,3-oxathiolanes

A uniform procedure was used for the photolyses of all substituted 1,3-dioxolanes and 1,3-oxathiolanes as exemplified for 2-cyclopropyl-1,3-dioxolane.

A mixture of 3.0 g of 2-cyclopropyl-1,3-dioxolane, 7.5 g of benzophenone and 200 ml of dry CFCl$_3$ was deoxygenated and irradiated for 17 hrs at 0°. The solvent was evaporated at atmospheric pressure (some HCl was also evolved) and the residue distilled under reduced pressure, yielding 1.95 g of the 2-chloroethyl ester of cyclopropane carboxylic acid boiling at 41–42°/4 mm. A second fraction with b.p. 124–125°/4 mm solidified and proved to be unreacted benzophenone. In some

---

[36] R. Leutner, *Monatsh. Chem.* **66**, 222 (1935).

[37] Commercial sample of 97.6% optical purity; a correction factor of 1.025 has therefore been applied in calculating specific rotations of derivatives.

[38] Values of 67–68°C and −40.6° are reported for the enantiomer; *cf.* V. Prelog, M. Wilhelm and D. Bruce Bright, *Helv. Chim. Acta* **37**, 221 (1954).

cases a small amount of precipitate was formed during the photolysis; this was shown to be benzpinacol.

The photolyses of the 1,3-oxathiolanes proceeded slower and some starting material was always recovered after the photolyses. The resulting (thio)esters were purified by GLC. All spectral data were in accordance with proposed structures (table 2). The *new* S-2-chloroethyl thioesters were also characterized by their microanalyses:

S-2-chloroethyl thiopropionate. Anal. Found: C, 39.31; H, 6.04; Cl, 23.06; S, 20.90; C_5H_9ClOS requires: C, 39.34; H, 5.94; Cl, 23.23; S, 21.01%.

S-2-chloroethyl thiobutyrate. Anal. Found: C, 43.46; H, 6.72; Cl, 21.15; S, 19.07; $C_6H_{11}ClOS$ requires: C, 43.24; H, 6.65; Cl, 21.27; S, 19.24%.

S-2-chloroethyl thio*iso*butyrate. Anal. Found: C, 43.17; H, 6.76; Cl, 21.24; S, 19.18; $C_6H_{11}ClOS$ requires: C, 43.24; H, 6.65; Cl, 21.27; S, 19.24%.

Hydrolysis of 2RS,4R-(−)-2-methyl-4-phenyl-1,3-dioxolane (X)

Compound X (0.6 g) was hydrolysed in a mixture of 30 ml of 10% H_2SO_4 and 5 ml of dioxane in 3 hrs at 50°. The solution was extracted three times with 40 ml portions of ether; drying (K_2CO_3) and evaporation of the solvent yielded 0.12 g of R-(−)-1-phenylethane-1,2-diol with $[\alpha]_D$ −40.2° (c, 1.2, 96% ethanol).

Photolysis of 2RS,4R-(−)-2-methyl-4-phenyl-1,3-dioxolane (X)

Compound X (4.3 g), benzophenone (8.0 g) and 160 ml $CFCl_3$ were irradiated under nitrogen during 24 hrs at 0°. After removal of the solvent, the residue was distilled under reduced pressure; the main fraction boiled at 102–105°/1 mm, yield 2.0 g. It was concluded from the NMR spectra that apart from unreacted X and benzophenone only two products were formed; the yields of *S-(+)-2-acetoxy-1-chloro-1-phenylethane* (XVII) and *R-(−)-1-acetoxy-2-chloro-1-phenylethane* (XVIII) were 52.6 and 4.0% respectively (ratio 93:7); *cf.* scheme 4 and table 2. Both XVII and XVIII are *new* compounds and were synthesized *via* independent routes to determine their specific rotations: $[\alpha]_{578}$ +97.1° and −77.3° respectively (in cyclohexane). The main fraction obtained by photolysis of X consisted of 5% of X, 73% of XVII, 5% of XVIII and 18% of benzophenone, as determined from the NMR spectrum. The observed rotation of this mixture at 578 nm was +5.60° (c, 8.55, cyclohexane, *l*, 1 dm). This rotation is the sum of the rotations of X, XVII and XVIII. Since we are only interested in the rotation of XVII we have to make corrections for the presence of small amounts of X and XVIII. These two however are present in their optical pure form (the former is starting material and chiral center C-4 is not involved in the formation of the latter) and their "sub-concentrations" can easily be calculated; the rotations at 578 nm are −0.28° and −0.33° respectively. The value for α_{578} is therefore +6.21° for XVII; the sub-concentration is 6.23, so $[\alpha]_{578}$ is +99.7° (cyclohexane). Since the specific rotation of optically pure XVII is $[\alpha]_{578}$ +97.1°, XVII is formed during the photolysis with 100% inversion of configuration.

Synthesis of S-(+)-2-acetoxy-1-chloro-1-phenylethane (XVII)

S-(+)-mandelic acid was treated with thionyl chloride according to McKenzie and Barrow[39] yielding S-(+)-phenylchloroacetic acid with $[\alpha]_{578}$ +113.5° (c, 0.8, benzene); the optical purity is therefore 60%.[40]

A solution of 2.0 g of B_2H_6 in 25 ml of THF was added at 0°C to a solution of 3.3 g of S-(+)-phenylchloroacetic acid in 40 ml of dry THF. The mixture was refluxed for 1 hr and poured into ice-water. After saturation with NaCl the mixture was extracted with ether; drying ($MgSO_4$) and evaporation of the solvent afforded 3.0 g of crude S-(+)-2-chloro-2-phenylethanol. The NMR spectrum did not reveal the presence of other compounds and the product was used without purification.

A solution of 8 ml of freshly distilled acetyl chloride in 8 ml of cyclopentane was added in 15 min. to a refluxing solution of 2.7 g of S-(+)-2-chloro-2-phenyl-ethanol in 10 ml of cyclopentane; the mixture was refluxed for another 5 hrs and was poured into an ice-cold $NaHCO_3$ solution. Extraction with ether, drying ($MgSO_4$), removal of the solvent and distillation afforded 2.1 g of S-(+)-2-*acetoxy-1-chloro-1-phenylethane*, b.p. 113–115°/2 mm, $[\alpha]_{578}$ +58.0° (c, 5.0, cyclo-hexane). For spectral data: *cf*. table 2. Anal. Found: C, 60.59; H, 5.46; Cl, 17.66; $C_{10}H_{11}ClO_2$ requires: C, 60.46; H, 5.58; Cl, 17.85%. Since the optical purity of the S-(+)-phenylchloroacetic acid was 60%, the specific rotation of optically pure XVII will be $[\alpha]_{578}$ +97.1°, $[\alpha]_D$ +93.0°.

Synthesis of R-(−)-1-acetoxy-2-chloro-1-phenylethane (XVIII)

Racemic 2-chloro-1-phenylethanol was prepared according to Sumrell *et al.*[41] A mixture of equimolar amounts of 2-chloro-1-phenylethanol, phthalic anhydride and pyridine was heated at 110° for 2 hrs; the mixture was poured into dilute HCl; extraction with chloroform, drying ($MgSO_4$) and evaporation of the solvent yielded the hydrogen phthalate of (racemic) 2-chloro-1-phenylethanol, which was crystallized from benzene-light petroleum 1:1, m.p. 91.5–92.5°; yield 87%.

The brucine salt of this hydrogen phthalate was resolved by fractional crystalli-zation from acetone. The less soluble salt was recrystallized five times until the specific rotation and melting point were constant: $[\alpha]_{364}$ −50.8° (c, 2.1, 96% ethanol), m.p. 126–128°, white needles. Treatment with dilute HCl, extraction with ether, washing of the ether solution with very dilute HCl, drying ($MgSO_4$) and removal of the solvent yielded the optically pure hydrogen phthalate of (−)-2-chloro-1-phenylethanol, obtained as a very viscous oil,[42] which was dextro-rotatory (in ethanol). Since it was not sure whether the solvent was removed completely, the calculated specific rotation might not be accurate: $[\alpha]_{364}$ +78.7° (c, 3.0, 96% ethanol).

The normal procedure to convert hydrogen phthalates of optically active alcohols into the alcohols without racemization is saponification with simultaneous steam distillation.[43] This procedure fails in our case, because only epoxyethyl-

---

[39] A. McKenzie and F. Barrow, *J. Chem. Soc.* 1910 (1911).

[40] A value of −190° is reported for the pure enantiomer; *cf*. K. Freudenberg, J. Todd and R. Seidler, *Liebigs Ann.* **501**, 199 (1933).

[41] G. Sumrell, B. M. Wyman, R. G. Howell and M. C. Harvey, *Can. J. Chem.* **42**, 2896 (1964).

[42] A sample of this oil was kept for several weeks at room temperature and solidified to a crystalline mass, m.p. 71–72°.

[43] R. H. Pickard and J. Kenyon, *J. Chem. Soc.* 45 (1911).

benzene is formed under the basic conditions. However, this reaction was favourably used to establish the absolute configuration of the hydrogen phthalate of (−)-2-chloro-1-phenylethanol and derivatives.

A solution of 0.5 g of NaOH in 30 ml of boiling water was added to 1.0 g of the hydrogen phthalate; steam distillation was carried out immediately, the distillate was extracted with ether, dried (MgSO$_4$) and the solvent evaporated to give 0.24 g of optically active epoxyethylbenzene, $[\alpha]_D +3.8°$ (c, 1.2, cyclohexane); (+)-epoxyethylbenzene is known to have the R-configuration.[44,45] The hydrogen phthalate of (−)-2-chloro-1-phenylethanol also has the R-configuration since R-(+)-epoxyethylbenzene is formed with retention of configuration in the saponification reaction.

The following method was used to convert the optically pure hydrogen phthalate into R-(−)-2-chloro-1-phenylethanol. A solution of 4.4 g of B$_2$H$_6$ in 55 ml of THF was added at 0° to a solution of 6.2 g of the hydrogen phthalate in 40 ml of dry THF. The mixture was refluxed for 1 hr and was then poured into ice-water. After saturation with NaCl the mixture was extracted with ether; drying (MgSO$_4$) and evaporation of the solvent afforded a residue consisting of only R-(−)-2-chloro-1-phenylethanol and phthalyl alcohol. The former was isolated by distillation, to yield 2.35 g, b.p. 119–120°/11 mm, $[\alpha]_D -47.8°$ (c, 2.8, cyclohexane); spectra of this *new* compound were identical with those obtained from racemic material.

A solution of 9 ml of freshly distilled acetyl chloride in 8 ml of cyclopentane was added to a refluxing solution of 2.0 g of R-(−)-2-chloro-1-phenylethanol in 20 ml of cyclopentane; the mixture was refluxed for 12 hrs. The excess of acetyl chloride and the solvent were evaporated, ether was added to the residue and the ether solution was washed with NaHCO$_3$ solution and finally with water; drying (MgSO$_4$), removal of the solvent and distillation produced 1.9 g of R-(−)-1-*acetoxy-2-chloro-1-phenylethane* (XVIII), b.p. 119°/1.5 mm, $[\alpha]_{578} -77.3°$, $[\alpha]_D$ $-73.6°$ (c, 2.9, cyclohexane). For spectral data: *cf.* table 2. Anal. Found: C, 60.58; H, 5.62; Cl, 18.02; C$_{10}$H$_{11}$ClO$_2$ requires: C, 60.46; H, 5.58; Cl, 17.85%.

ESR of 1,3-*dioxolanes*

Optimal results are obtained when equal amounts of a substituted 1,3-dioxolane and di-t-butyl peroxyoxalate[13] (TBPO) are dissolved in a dilute solution of 2-nitroso-2-methylpropane (t-BuNO) in benzene (roughly 5 mg per ml). TBPO decomposes slowly at room temperature and immediately the nitroxide (VI) derived from the trapped acetal radical is observed; sometimes a small signal due to di-t-butyl nitroxide ($a_N = 15.5$ gauss) is also present. Photolysis of the same sample produces large amounts of this nitroxide, which obscures the signal derived from VI. However, in some cases the nitroxides derived from the cyclic acetal radicals (VI) could be observed in photochemical benzophenone-induced reactions using very short (10 sec.) irradiation times.

[44] I. Tömöskösi, *Tetrahedron* **19**, 1969 (1963).
[45] C. R. Johnson and C. W. Schroeck, *J. Amer. Chem. Soc.* **90**, 6852 (1968).

ESR of 1,3-*oxathiolanes*

Similar results are obtained with 1,3-oxathiolanes using the TBPO method. Nitroxides (XXII) derived from the trapped thioacetal radicals are observed; sometimes small signals due to (unidentified) impurities are also present in the spectra. No radicals of type XXII could be detected during brief photolysis in a benzophenone-induced reaction; only di-*t*-butyl nitroxide was observed. Presumably photochemically induced abstraction of the hydrogen at C-2 is a more difficult process in substituted 1,3-oxathiolanes as compared with substituted 1,3-dioxolanes.

SUMMARY

This thesis is based on the excellent radical scavenging properties of monomeric nitrosoalkanes RNO. Thus, addition of a small amount of RNO to a reacting system in which unstable radicals X· are generated usually results in the formation of relatively stable nitroxides, which can be studied by ESR spectroscopy at room temperature.

$$R-N=O \ + \ X^{\bullet} \longrightarrow \underset{\underset{O}{\overset{\bullet}{|}}}{R-N-X}$$

The nature of the trapped radical X· can very often be deduced from the hyperfine splittings (hfs) and g-value of the ESR spectrum of the derived nitroxide. Thus the ESR Spin Trapping technique is useful for the identification of highly reactive (short-living) radicals X·.

Chapter 1 contains a survey of the literature and a short description of the most important features of nitrosoalkanes, nitroxides and the Spin Trapping technique in general.

In chapter 2 the synthesis of hitherto unknown 1-nitrosoadamantane is described. Together with a number of other nitrosoalkanes it is used as a scavenger for trifluoromethyl radicals, generated by photolysis of trifluoro-iodomethane. The derived alkyl trifluoromethyl nitroxides with $a_N = a_F = $ 11.5–12.0 gauss are easily identified and are found to be very stable at room temperature.

Chapter 3 illustrates the limitations of the Spin Trapping technique, $i.e.$ the nitroxides formed when a nitrosoalkane is added to a system in which certain radicals are involved, are not necessarily the addition products from these radicals. The primarily formed nitroxides are sometimes unstable and secondary reactions may ultimately produce new nitroxides, only indirectly related to the initially formed radicals.

This is found to be case when various polyhalomethanes (except precursors of CF_3 radicals) are photolysed in the presence of 2-nitroso-2-methyl-

45

propane. The initially generated tri-, di- and monohalomethyl radicals form adducts with the scavenger, but in sharp contrast with the fluorine-containing nitroxides discussed in chapter 2, these are very unstable when containing the other halogens (Cl, Br or I). By elimination of (formally atomic) halogen they produce halonitrones. These (halo)nitrones ultimately afford the stable acyl alkyl nitroxides ($a_N = 6.5$–7.0 gauss) that are the only radicals recorded under the photochemical conditions. A mechanism is proposed in which the halonitrones undergo light-induced isomerization to halo-oxaziridines; in a bimolecular oxidation-reduction reaction these can lead to the formation of the observed acyl alkyl nitroxides.

Chapter 4 illustrates the use of Spin Trapping in studying hydrogen abstraction from substituted 1,3-dioxolanes. The resulting cyclic acetal radicals are trapped efficiently by 2-nitroso-2-methylpropane (t-BuNO).

Surprisingly, no ring-opened primary ester radicals are trapped at all, although in quantitative experiments ring-opened products are often isolated in high yields. For instance when the photolysis of 2-alkyl-1,3-dioxolanes is carried out in $CFCl_3$ in the presence of benzophenone open esters $RCOOCH_2CH_2Cl$ are obtained. The results of the photolysis of optically

46

active $2RS,4R$-$(-)$-2-methyl-4-phenyl-1,3-dioxolane strongly suggest that these esters are not formed *via* ring-opened ester radicals, but *via* an ionic mechanism, in full agreement with the ESR spin trapping experiments which indicated the absence of ring-opened radicals. The main product, S-$(+)$-2-acetoxy-1-chloro-1-phenylethane, is formed with 100% inversion of configuration and this is only compatible with a mechanism in which an S_N2 reaction is involved (*vide infra*). A mechanism involving homolytic ring-opening of the cyclic acetal radical seems less likely since it would result, at least partly, in the formation of racemic product.

A similar study on 2-alkyl-1,3-oxathiolanes is also presented in chapter 4. Again, only *cyclic* thioacetal radicals are trapped when 2-nitroso-2-methyl-propane is used as a scavenger. When the photolysis is carried out in $CFCl_3$ in the presence of benzophenone only the S-2-chloroethyl thioesters are obtained and no trace of the isomeric O-2-chloroethyl thioesters. This selectivity is well understood in terms of the ionic mechanism.

The fact that a carbonium ion is much better stabilized by a neighbouring oxygen atom than by sulfur, renders the (other) carbon atom attached to oxygen more susceptible for nucleophilic attack by the chloride ion.

47

SAMENVATTING

De studie beschreven in dit proefschrift is gebaseerd op de eigenschap van monomere nitrosoalkanen RNO om efficiënt vrije radicalen weg te vangen. Wanneer een kleine hoeveelheid RNO wordt toegevoegd aan een reagerend systeem waarin kort levende radicalen X· als intermediairen een rol spelen, vindt in vele gevallen een additie-reactie plaats onder vorming van stabiele nitroxiden, die m.b.v. ESR spectroscopie bij kamertemperatuur bestudeerd kunnen worden.

$$\text{R-N=O} \;+\; \overset{\displaystyle\bullet}{\text{X}} \;\longrightarrow\; \underset{\displaystyle \text{O}}{\overset{\displaystyle |\bullet}{\text{R-N-X}}}$$

De structuur van het weggevangen radicaal X· kan dikwijls worden afgeleid uit de hyperfijnsplitsingen (hfs) en de g-waarde van het ESR spectrum van het gedetecteerde nitroxide. De ESR *Spin Trapping* techniek is daarom een nuttige methode om zeer reactieve (kort levende) radicalen X· te identificeren.

Hoofdstuk 1 bevat een overzicht van de literatuur en een korte beschrijving van de belangrijkste eigenschappen van nitrosoalkanen en nitroxiden, terwijl de *Spin Trapping* techniek kort wordt toegelicht.

In hoofdstuk 2 wordt de synthese van 1-nitrosoadamantaan beschreven. Met een aantal reeds bekende nitrosoalkanen wordt deze nitrosoverbinding gebruikt om trifluormethyl radicalen (verkregen door fotolyse van trifluorjoodmethaan) weg te vangen. De gevormde alkyl trifluormethyl nitroxiden, die zeer stabiel blijken te zijn bij kamertemperatuur, kunnen gemakkelijk aan de hand van hun ESR spectra ($a_N = a_F = 11.5–12.0$ gauss) geïdentificeerd worden.

Hoofdstuk 3 beschrijft de beperkingen van de *Spin Trapping* techniek. Het blijkt dat de nitroxiden, die worden waargenomen bij toevoeging van een nitrosoalkaan aan een radicalen genererend systeem, niet noodzakelijkerwijs de additieproducten van die radicalen behoeven te zijn. De primair gevormde nitroxiden zijn soms instabiel en secundaire reacties kunnen aanleiding geven tot de vorming van nieuwe nitroxiden die slechts indirect in verband staan met de oorspronkelijk gevormde radicalen.

Dergelijke complicaties treden op wanneer polyhalomethanen (behalve bronnen van CF_3 radicalen) worden bestraald in aanwezigheid van 2-nitroso-2-methylpropaan. De primair gevormde tri-, di- en monohalomethyl radicalen vormen adducten met de nitrosoverbinding maar deze zijn, in scherpe tegenstelling tot de fluor bevattende nitroxiden uit hoofdstuk 2, zeer instabiel wanneer ze andere halogenen (Cl, Br of I) bevatten. Door eliminatie van een halogeenatoom worden halonitronen gevormd. Deze halonitronen leveren uiteindelijk relatief stabiele acyl alkyl nitroxiden met $a_N = 6.5–7.0$ gauss; dit zijn de enige radicalen die tijdens de bestraling worden waargenomen. Een mechanisme wordt voorgesteld waarin de halo-nitronen onder invloed van licht isomeriseren tot halo-oxaziridinen. Deze kunnen tenslotte de waargenomen acyl alkyl nitroxiden leveren via een bimoleculaire oxidatie-reductie reactie.

In hoofdstuk 4 wordt *Spin Trapping* gebruikt om de waterstofabstractie van gesubstitueerde 1,3-dioxolanen te bestuderen. De resulterende cyclische acetaal radicalen worden efficiënt weggevangen door 2-nitroso-2-methyl-propaan (t-BuNO).

49

Het is verrassend dat er geen ESR spectra worden waargenomen van nitroxiden afgeleid van opengesplitste primaire ester radicalen, aangezien er in kwantitatieve experimenten hoge opbrengsten aan acyclische producten (ontstaan door ringopening van de dioxolaanring) worden geïsoleerd. Zo worden er bijvoorbeeld open esters $RCOOCH_2CH_2Cl$ verkregen wanneer de fotolyse van 2-alkyl-1,3-dioxolanen wordt uitgevoerd in $CFCl_3$ in aanwezigheid van benzophenon. Deze esters worden echter niet gevormd via open ester radicalen $RCOOCH_2CH_2\cdot$, maar via een ionogeen mechanisme, zoals blijkt uit het stereochemisch verloop van de fotolyse van optisch actief $2RS,4R$-(−)-2-methyl-4-phenyl-1,3-dioxolaan. Dit resultaat is geheel in overeenstemming met de ESR experimenten, die de afwezigheid van opengesplitste ester radicalen aantoonden. De vorming van het hoofdproduct, S-(+)-2-acetoxy-1-chloor-1-phenylethaan (met 100% inversie van configuratie) is alleen te rijmen met een mechanisme waarin een S_N2 reactie een rol speelt (*vide infra*). Een mechanisme waarbij homolytische opening van het cyclische acetaal radicaal optreedt, lijkt onwaarschijnlijk omdat dit aanleiding zou geven tot de vorming van, tenminste gedeeltelijk, racemisch product.

Een soortgelijke studie over 2-alkyl-1,3-oxathiolanen wordt eveneens beschreven in hoofdstuk 4. Ook hier worden alleen *cyclische* thioacetaal radicalen weggevangen wanneer 2-nitroso-2-methylpropaan wordt gebruikt. Indien de fotolyses worden uitgevoerd in $CFCl_3$ in aanwezigheid van benzophenon, worden er alleen S-2-chloorethyl thioesters geïsoleerd en geen spoor van de isomere O-2-chloorethyl thioesters. Deze selectiviteit is goed te verklaren m.b.v. het ionogene mechanisme.

$$H \diagdown \; O-CH_2 \qquad \xrightarrow{\text{3 steps}} \qquad Cl^{\ominus} \; O-CH_2 \qquad \longleftrightarrow \qquad Cl^{\ominus} \qquad O-CH_2 \qquad \longrightarrow \qquad O$$

Het feit dat een carbonium ion veel beter door een naburig zuurstofatoom wordt gestabiliseerd dan door zwavel, maakt het (andere) koolstofatoom naast zuurstof kwetsbaar voor een nucleofiele aanval van het chloride ion.